UML-B SPECIFICATION FOR PROVEN EMBEDDED SYSTEMS DESIGN

UML-B Specification for Proven Embedded Systems Design

by:

Fredrik Bernin, *Volvo Technology Corporation, Sweden*
Michael Butler, *University of Southampton, Great-Britain*
Dominique Cansell, *LORIA, France*
Stefan Hallerstede, *KeesDA S.A., France*
Klaus Kronlöf, *Nokia Research Center, Finland*
Alexander Krupp, *Paderborn University/C-LAB, Germany*
Thierry Lecomte, *ClearSy, France*
Michael Lundell, *Volvo Technology Corporation, Sweden*
Ola Lundkvist, *Volvo Technology Corporation, Sweden*
Michele Marchetti, *Nokia Research Center, Finland*
Wolfgang Mueller, *Paderborn University/C-LAB, Germany*
Ian Oliver, *Nokia Research Center, Finland*
Denis Sabatier, *ClearSy, France*
Tim Schattkowsky, *Paderborn University/C-LAB, Germany*
Colin Snook, *University of Southampton, Great-Britain*
Nikolaos S. Voros, *INTRACOM S.A., Greece*
Yann Zimmermann, *KeesDA S.A., France*

Edited by

Jean Mermet, *KeesDA S.A., France*

KLUWER ACADEMIC PUBLISHERS
BOSTON / DORDRECHT / LONDON

A C.I.P. Catalogue record for this book is available from the Library of Congress.

ISBN 978-1-4419-5256-1 (**Pb**) e-ISBN 978-1-4020-2867-0 (e-book)

Published by Kluwer Academic Publishers,
P.O. Box 17, 3300 AA Dordrecht, The Netherlands.

Sold and distributed in North, Central and South America
by Kluwer Academic Publishers,
101 Philip Drive, Norwell, MA 02061, U.S.A.

In all other countries, sold and distributed
by Kluwer Academic Publishers,
P.O. Box 322, 3300 AH Dordrecht, The Netherlands.

Printed on acid-free paper

All Rights Reserved
© 2010 Kluwer Academic Publishers, Boston
No part of this work may be reproduced, stored in a retrieval system, or transmitted
in any form or by any means, electronic, mechanical, photocopying, microfilming, recording
or otherwise, without written permission from the Publisher, with the exception
of any material supplied specifically for the purpose of being entered
and executed on a computer system, for exclusive use by the purchaser of the work.

Printed in the Netherlands.

Contents

Preface

This book presents the perspective of the project on a **P**aradigm **U**nifying **S**ystem **S**pecification **E**nvironments for proven **E**lectronic design (PUSSEE) as conceived in the course of the research during 2002 - 2003.

The initial statement of the research was formulated as follows:

The objective of PUSSEE is to introduce the formal proof of system properties throughout a modular system design methodology that integrates sub-systems co-verification with system refinement and reusability of virtual system components. This will be done by combining the UML and B languages to allow the verification of system specifications through the composition of proven sub-systems (in particular interfaces, using the VSIA/SLIF standard). The link of B with C, VHDL and SystemC will extend the correct-by-construction design process to lower system-on-chip (SoC) development stages. Prototype tools will be developed for the code generation from UML and B, and existing B verification tools will be extended to support IP reuse, according to the VSI Alliance work. The methodology and tools will be validated through the development of three industrial applications: a wireless mobile terminal–a telecom system-on-chip based on HIPERLAN/2 protocol and an anti-collision module for automobiles.

The problem was known to be hard and the scope ambitious. But the seventeen chapters that follow, describing the main results obtained demonstrate the success of the research, acknowledged by the European reviewers. They are released to allow the largest audience to learn and take benefit of.

Acknowledgements

The research work that provided the material for this book was carried out during 2002 - 2003 mainly in the PUSSEE project, supported partially by the European Commission under the contract IST-2000-30103. Guidance and comments of G. Schumacher, R. Reed and J. Khan on research direction are highly appreciated.

In addition to the authors, the contributions of the following project members and partners' personnel are gratefully acknowledged: F.Pourchelle, K. Sandström, S. Batistatos, K. Tibas, C. Drosos, K. Antonis, S. Kostavasili, C. Dre, S. Sakelliou, J. Karathanasis, S. Deirmentzoglou, L. Voisin, E. Jaffuel, J.P Pitzalis, J.R. Abrial, C. Metayer, J.L. Vergely, C. Czernecki, L. Draper, P. Wheeler, E. Turner, P. Turner, C. Loeser, S. Flake and J. Ruf.

Chapter 1

AN INTRODUCTION TO FORMAL METHODS

How do they apply in embedded system design?

Nikolaos S. Voros[1]
Wolfgang Mueller[2]
Colin Snook[3]

[1] INTRACOM S.A., Patra, Greece
[2] Paderborn University/C-LAB, Paderborn, Germany
[3]University of Southampton, Southampton, United Kingdom

Abstract: This chapter begins with an introduction to the main concepts of formal methods. Languages and tools for developing formal system models are also described, while the use of semi formal notations and their integration with formal methods is covered as well. At the end of the chapter, an overview of the current status of formal methods in embedded system design is presented.

Key words: Formal methods, formal languages, semi formal notations, tools for formal languages, embedded system design.

1. INTRODUCTION

Since we focus our investigations on formal specification and verification, we first give definitions of basic terms before introducing the theoretical background, formal languages and tools.

A *specification* can be regarded as a description that is intended to be as *precise*, *unambiguous*, *concise and complete* as possible in the context of its specific application [1]. A *formal specification* is a specification written in a formal language where a *formal language* is either based on a rigorous mathematical model or simply on a standardised programming or

J. Mermet (ed.), UML-B Specification for Proven Embedded System Design, 1–20.
© 2004 *Kluwer Academic Publishers. Printed in the Netherlands.*

specification language [2]. Due to its individual application, a formal specification can be (partly) *executable.* In most cases, formal specifications are for a mental execution by code review and for passing the specification around to members in a design team. In most cases only subsets of formal specification languages, e.g. of Z and VDM, are machine executable. A *formal method* implies the application of at least one formal specification language. Formal methods are often employed during system design when the degree of confidence in the prescribed system behaviour, extrapolated from a finite number of tests, is low. Formal methods are frequently applied in the design of ultra-reliable as well as complex concurrent or reactive systems. Formal specifications can be classified with respect to their specification style. Here, we can identify mainly two different classifications. One is mainly due to the field of programming languages; the other one comes from general systems specification.

1.1 Axiomatic vs. Denotational vs. Operational

In the context of the formal semantics of programming languages, we distinguish axiomatic, denotational and operational styles [3]. For *axiomatic semantics,* each language construct is specified in terms of axioms and deduction rules, where the Hoare style is a frequently applied notation [4]. *Denotational semantics* is a sort of functional semantics. Specification of denotational semantics consists of two main steps [5]. First, the syntactic domain, the syntactic constructs and the semantic domains are identified. Secondly, the semantic (valuation) functions mapping the syntax of a language to its semantics including its signatures are defined in detail. Valuation functions are usually given by a set of equations in form of typed λ-calculus notation [6]. *Operational semantics* finally defines a set of explicit commands changing the state of a more or less abstract machine. Operational semantics is usually close to the implementation of a language. Classical means for the definition of operational semantics are Finite State Machines, programming language-oriented notations, pseudo code and related means.

1.2 Model-based vs. Algebraic

Sommerville divides specification styles for non-trivial systems in *model-based* and *algebraic* approaches [7]. The model-based style gives the explicit specification of abstract machines. For a model-based specification, a system is defined in terms of mathematic entities (e.g. sets, relations, sequences). Operations are defined in terms of abstract states and how they affect those entities. The algebraic style specifies abstract data types in terms of axioms defining relationships between its operations. This style is sometimes also denoted as *property-oriented* specification.

Due to recent trends in formal methods there is a huge number of formal methods, languages and available tools, each developed for a particular domain and purpose. It is impossible to give a complete enumeration of all known formal languages and tools for formal verification. We try to give a representative overview of widelly accepted and frequently applied approaches for formal specification and verification in the next sections. We first explore languages based on predicate logic, temporal logic, process algebras before we outline means, which do not fit in just one category like RSL, Action Semantics and Abstract State Machines. The final overview of formal verification tools mainly investigates model checkers and theorem provers.

1.3 Theory

A huge array of formal specification means and verification tools is based on predicate logic so that we start with an overview of the theoretical background on predicate logic. *Predicate logic* is a branch of mathematical logic investigating the interpretation, construction and derivation of well-formed formulae (WFF). In first-order predicate logic, an atom has the form $P(t_1, ..., t_k)$, where P is a predicate symbol and $t_1, ..., t_k$ are terms, which can be composed of function symbols, constant symbols and variables. *High-order predicate logic* is defined by additionally allowing the application of predicate symbols in terms. A well-formed formula is recursively composed of atoms, formulae, propositions or connectives (negation, disjunction, conjunction, implication, equivalence), conditions (equal, less, greater) and (universal and existential) quantifiers. Predicate logic is typed and multi-sorted. Predicate logic without predicates, functions and quantifiers is traditionally called *propositional logic*. In the context of predicate logic, we can distinguish model theory and proof theory [8].

Model theory investigates the interpretation of syntactical sentences with respect to a model. The model is given by objects of a chosen domain. Syntactical sentences are composed of symbols, which are associated with objects by defining truth-valued functions that give the required interpretation. In predicate logic, well-formed formulae are composed of constant, predicate; function symbols are typically interpreted by Boolean-valued functions. A well-formed formula is *satisfiable* if there exists one interpretation that evaluates to true. An *axiom* is a well-formed formula that is true by definition. *Modal logic* considers the interpretation of formulae within different contexts (*worlds*) by so-called modal operators, such as *necessity* and *possibility*. *Temporal logic* is an instance of modal logic with respect to time modalities [9]. Time modalities are defined by temporal future operators (e.g. Next, Henceforth, Eventually, Until, Unless) and past operators (e.g. Previous, Has-Always-Been, Once, Since, Back-To) [10]. Temporal operators typically refer to sequences of state transitions. Future-oriented temporal logic distinguishes *branching-time* and *linear-time* temporal logic. Branching-time temporal logic considers all possible paths within the tree defined by the set of state transitions, where the root of the tree is given by the current state. Linear-time temporal logic investigates just one path of that tree.

Proof theory investigates the relationship between sentences of a formal system using only rules for operating on the syntactical content of those sentences. A relation transforming a well-formed formula to another well-formed formula is denoted as an *inference rule*. An inference rule is composed of a set of sequents $S_1, ..., S_r$ and another sequent S, which follows from that set[1]:

$$\begin{array}{|c|} S_1 \\ \vdots \\ S_r \\ \hline S \end{array}$$

A *sequent* is a pair whose first component is a set of formulae $A_1, ..., A_n$ (hypothesis) and its second component is a formula B (conclusion) that follows from the hypothesis: $A_1, ..., A_n \vdash B$

An example of an inference rule is the *modus ponens* which defines that if A holds and A implies B then B follows:

$$\begin{array}{|c|} \vdash A \\ \vdash A \Rightarrow B \\ \hline \vdash B \end{array}$$

The application of inference rules is called a deduction. A deduction can be represented as a sequence $f_1, ..., f_n$ of formulae, where each f_i, $2 \leq i \leq n$, is deduced from f_{i-1} by the application of an inference rule. f_1 is an axiom and f_n is typically denoted as a theorem. A set of axioms and the set of all theorems derivable from the axioms by the application of all inference rules is called a *theory*. A theory is consistent if, and only if, it does not contain both *s*

[1] Sometimes also written as $\{S_1, ..., S_r\} \vdash S$

and $\neg s$ for all sentences s in the theory. Theorem provers are typically based on the sequent calculus and compute a deduction for a given set of axioms, theorems and inference rules.

A formal system of previous form is *decidable* if an algorithm can be constructed, which decides whether a given well-formed formula is a theorem. Propositional logic is decidable; first-order (and high-order) predicate logic is not decidable. When interpreting well-formed formulae, a formal system is *complete* if all true formulae can be derived. A formal system is *sound* if all derivable formulae are true. Propositional and first-order logic are sound and complete. High-order logic is sound and not complete.

2. LANGUAGES

To give a overview of formal specification languages, we have selected a representative set of languages from predicate logic, temporal logic, process algebras and a set of languages which do not fit into the previous categories.

2.1 Predicate Logic-Based Languages

In predicate-logic based specification languages, a system is usually defined by a set of functions, where each function is defined by its signature and a set of predicates that define pre- and post-conditions for the function. We introduce two classical predicate-logic based languages, Z and VDM, and the B language.

2.1.1 Z

Z was developed by the Programming Research Group at Oxford University and accepted as a BSI standard in 1989 [11]. Z is a specification language based on set theory with no official method. Object-Oriented and real-time extensions to Z are available as Object-Z and Timed Communicating Object-Z, respectively. There are conventions and practices to use Z as a model-based language. Though Z permits various specification styles, the state-based approach has been found the most convenient one in many applications. Z comes with a deductive system in order to reason about specifications. Z is based on typed set theory and first-order predicate logic. Invariants can be associated with a global state. Invariants relate pre- and post- conditions for operations. A Z specification is composed of modules (*schemas*). A Z schema is given by a schema name followed by its signature (set of domains) and a set of properties. Schemas support the separation of the specification (and error conditions) from the system's behaviour. Schemas are combined by a schema calculus. For a model-based specification, the domains are typically defined by basic sets and abstract states are defined in terms of sets, relations, functions, sequences, etc. Properties are assigned to states by means of extra predicates, e.g. initial conditions for the initial state. Abstract operations define state transitions.

The wider acceptance of Z in recent years has advanced the available set of tools. Tools for assistance with proof, and tools for translating Z specifications to programming languages, are available. However, a first high-level Z specification is usually not machine-executable.

2.1.2 VDM

The Vienna Development Method (VDM) is a formal specification method with the model-based specification language VDM-SL (VDM Specification Language) [12]. VDM was initially developed for the formal description of PL/I at the IBM laboratory in Vienna. The VDM method considers the verification of step-wise refinement in the systems development process, i.e. data refinement and operation decomposition. VDM-SL is conceptually similar to Z. VDM specifications are based on logic assertions of abstract states (mathematic abstraction and interface specification). In contrast to Z, VDM uses keywords in order to distinguish the roles of different components while these structures are not explicit in Z. As with Z specifications, VDM specifications are usually not machine-executable. VDM supports the specification process by a mental execution with paper and pencil. However, proof assistance tools and tools for executing subsets are available.

2.1.3 B

B stands for a language, a method and tools [13]. The B language can considered to be a combination of Z and Pacal with some extensions for refinement. The B method is based on a hierarchical stepwise refinement and decomposition of a problem. After initial informal specification of requirements, an abstraction is made to capture, in a first formal specification, the most essential properties of a system. For example, these could be the main safety properties in a safety critical system. This top-level abstract specification is made more concrete and more detailed in steps, which may be one of two types. The specification can be refined either by changing the data structures used to represent the state information and/or by changing the bodies of the operations that act upon these data structures. Alternatively, the specification can be decomposed into subsections by writing an implementation step that binds the previous refinement to one or more abstract machines representing the interfaces of the subsections. In a typical B project, many levels of refinement and decomposition are used to fully specify the requirements. Once a stage is reached when all the requirements have been expressed formally, further refinement and decomposition steps add implementation decisions until a level of detail is reached at level B_0, where code can be automatically generated for Ada and C++. B processing tools, like Atelier-B from Clearsy, are advanced theorem provers with code generation, which automatically provide theorems, i.e. proof obligations.

2.2 Temporal Logic-Based Languages

The next two paragraphs present temporal logic-based approaches: *Computational Tree Logic* and *Temporal Logic of Action*. The first one was developed in the context of classical model checking for hardware verification. The latter one was initially developed in the context of imperative programming languages.

2.2.1 CTL

CTL (Computational Tree Logic) was defined as a branching-time temporal logic for model checking. Several variations of CTLs are known for practical applications: *CTL*, *ACTL* and *CTL**. All CTLs are future-oriented. Only some approaches extend CTL with past modalities. CTL formulae express information about states or state transitions [14]. A *path* defines one possible future execution over state transitions beginning at the current state. All possible execution paths establish a tree with the current state at its root. The temporal operators are composed of two parts. The first part defines the quantification (path quantifier). It specifies either the truth of all paths starting from the current state (**A**) or specifies the existence of a path with certain properties (**E**). The second part specifies the ordering of events along one path or a set of paths by the operators: next-time (**X**), until (**U**) releases (**V**). *CTL** is a superset of *CTL* allowing the more general application of path quantifiers (**A** and **E**) in path formulae. *ACTL* is a subset of CTL eliminating the existence quantifier **E**. ACTL was defined to make the model checking more efficient. Linear-time temporal logic (LTL) investigates just one path in the tree rather than the complete tree.

2.2.2 TLA

TLA (Temporal Logic of Actions) is temporal logic-based theory providing a logic for specifying and reasoning about concurrent and reactive systems [15]. TLA+ is the language for writing TLA specifications. The corresponding tool for mechanically checking TLA proofs is TLP (TL Prover), which is based on the Larch theorem Prover (LP) and a BDD-based model checker. TLA supports the specification of refinements and checks properties like fairness. A TLA specification is a list of formulae. A TLA formula is constructed by negation, conjunction, disjunction, implication and equivalence including the existential quantifier and predicates.

2.3 Process Algebras

Another important class of formal specification languages is based on *Process Algebras*. A description based on process algebra generally defines a set of processes. A process p is sensitive on a set of events E_p and/or performs some actions A. In process algebras, a process p is identified by its set of partially order events, i.e. its *trace*. A *process algebra* is defined on processes and (mainly) compositional operators on processes, such as alternative, sequential and parallel compositioning. Some process algebras have labelling functions, others have events or actions.

Process algebras can also be considered as labelled transition systems, where a process p is transformed to a new process p' after accepting an event/action [16]. A transition system can be represented by a graph whose nodes are processes and edges are given by partially ordered state transitions. Edges are labelled by events/actions. Some approaches are based on the notion of *traces*. A *trace* is a sequence of actions/events reflecting the history of state transitions of a process. The behaviour of a process can be partially characterised by a set of traces. A large set of process algebras are derived from Milner's CCS (Calculus of Communicating Systems). A well-known ancestor is Hoare's CSP (Communicating Sequential Processes). Several parallel programming and specification languages, such as OCCAM and the ISO standard LOTOS are based on CSP, for instance. In the next paragraphs, we additionally investigate Circal, which extends CCS by some features for hardware verification.

2.3.1 CSS

CCS (Calculus of Communicating Systems) [17] specifies a system as a set of asynchronously running processes performing, possibly non-deterministic, actions. CCS allows processes to be guarded by actions (*action-prefixing*). Processes are called agents in CCS and actions are referred to as the ports of an agent. Ports can be parameterised by free variables. Ports are divided in in- and out- port. Agents communicate by handshake via ports and are defined in terms of equations with a so-called agent expression on their right-hand side. Basic operators on agents are: prefixing, summation, composition, restriction, relabelling, simultaneous substitution and recursion. τ denotes the silent action and *1* the unit action. A set of equivalence relation supports the comparison of traces. A process logic PL (modal logic) covers the specification of properties over the set of traces in terms of formulae.

2.3.2 CSP

CSP (Communicating Sequential Processes) is conceptually similar to CCS [18]. CSP specifies a system as a set of asynchronously running processes acting on events. Processes communicate values (resp. events) via channels. To check CSP specification for properties, a satisfiability relation *sat* was introduced that allows CSP specifications to be checked with respect to a set of traces, i.e. specification vs. implementation. Hoare gives a listing of the differences between CCS and CSP in [18]. The main differences refer to the definition of non-determinism and the included correctness calculus. In CCS, non-determinism defined through τ is not associative in contrast to CSP. A major difference between CCS and CSP is in the correctness calculus. CCS defines a set of equivalences for observational behaviour, where CSP defines a satisfiability relation for specifying whether a process meets a given specification. CSP does not proof that an implementation does not satisfy a specification.

2.3.3 Circal

Circal (CIRcuit CALculus) is a process algebra for the formal verification of digital hardware including asynchronous hardware [19]. Circal defines a set of core operators and a set of derived laws. The laws are based on the semantics of the core operators using a labelled transition system and equivalence relations. Circal compares to CCS. In contrast to CCS (and CSP), Circal supports simultaneous actions and multiway composition. Circal allows processes to have guards with more than one action forming a *simultaneous guard,* which requires that all actions in the guard must occur simultaneously for the process to evolve. The Circal composition operator defines broadcast by allowing many processes to communicate via interconnected ports sharing a common label. This supports the modelling of bus structures. The combination of simultaneous actions and multiway composition provides means for describing synchronous systems through the inclusion of a clock action in a simultaneous guard.

2.4 Other Approaches

We finally sketch three prominent representatives of approaches which do not exactly fit into the previous categories: RSL, Action Semantics and Abstract State Machines. They all integrate functional, imperative and concurrent aspects.

2.4.1 RSL

RSL Specification Language is a "wide-spectrum" specification and implementation language. It supports abstract, property-oriented specifications as well as low-level designs with concrete algorithms supporting specification refinement and reuse [20]. RSL is a combination of several specification languages, i.e. VDM, CSP, ACT-ONE and Standard ML. The basic concepts are inherited from VDM: type constructors for mapping, sets, cartesian products and function definitions (pre- and post-style). Concepts of CSP (resp. CCS) are incorporated to manage concurrency (channels, communication, concurrent composition). Class expressions, schemes and objects are the basic modules in RSL and similar to Standard ML functors. Basic class expressions are defined by theories (signatures and axioms). The algebraic specification, where types and values are constrained by the means of axioms, is based on the concepts of ACT-ONE.

2.4.2 Action Semantics

Action Semantics (AS) is a framework for the specification of the formal semantics of programming languages [21]. AS incorporates algebraic, functional (denotational) and imperative means. AS is a pragmatic approach based on the idea of functional semantics. An AS specification covers the specification of syntactic entities, semantic entities and semantic functions mapping syntactic entities to semantic entities. For semantic entities AS has *actions, data and yielders*. Data are static entities. Yielders represent dynamic data (unevaluated items of data) whose value depends on the current information. Actions are dynamic computational entities that define the behaviour. An action may diverge, complete, escape or fail. An action can be non-deterministic. Action notation supports the specification of control-flow, data-flow, process activation and communication. Non-commercial tools are available for interpreting and checking AS specifications based on the principles of term rewriting (ASD Tools), as well as a AS to C compiler.

2.4.3 ASMs

Gurevich initially introduced Abstract State Machines (ASMs) as Evolving Algebras in 1991. A revised definition of with various extensions, commonly known as the Lipari Guide, was published in 1994 [22]. Whereas Gurevich has originally defined ASMs for considerations in complexity theory, multiple publications have demonstrated the applicability of ASMs for formal specification in various fields: hardware and software architecture, protocols, as well as programming and hardware description languages [23].

An ASM specification is a program executed on an abstract machine. The program comes in the form of rules. Rules are nested if–then–else clauses with a set of function updates in their body. Based on those rules, an abstract machine performs state transitions with algebras as states. Firing a set of rules in one step performs a state transition. ASMs are multi-sorted based on the notion of universes. Microsoft Research provides an executable version of ASMs, which execute ASMs under the .NET platform, i.e. AsmL [23].

3. TOOLS

Formal verification tools can roughly be classified in *equivalence checker*, *model checker* and *theorem prover* [14].

An *equivalence checker* compares two models for equivalence by applying different heuristics like reachability analysis, BBD equivalence or SAT-solving [14].

A *model checker* verifies a system against specified properties. The system is typically given as a collection of finite state machines. The property is usually defined in terms of temporal logic formulae [24].

For *theorem proving* the verification problem is given as a set of formulae and a set of inference rules. A theorem is then deduced from a set of axioms applying those rules. Theorem proving generally requires continuous user interactions where in most cases equivalence and model checking perform completely automatic verifications. The following gives a short overview of some model checkers and theorem provers.

3.1 Model Checkers

Model checking verifies a given set of state machines with respect to a set of temporal formulae, e.g. CTL or LTL. Many heuristics are applied to overcome the "state explosion" problem. We basically can distinguish explicit state checking and symbolic model checking. Explicit state checking is based on explicit state exploitation based on various breadth-first search and depth-first search heuristics. FDR, Spin and Murphi, are representatives of explicit state model checkers. Symbolic model checking is mostly based on the symbolic state representation by Binary Decision Diagrams (BDDs) [25] for efficient computation. BDD-based model checkers typically verify specifications with up to 200 state variables. Single systems with up to 10^{1300} reachable states (64 register ALU) have already been checked by the use of very dedicated abstraction techniques [24]. SMV is the classical BDD-based symbolic model checker, which many model checkers are based on, like the RuleBase from IBM and FormalCheck from Cadence. The following gives a short overview of a symbolic and a explicit state model checker, SMV and FDR.

SMV (Symbolic Model Verifier) from CMU [26] was the first model checker based on BDDs. The classical SMV checks finite state systems against CTL specifications using a OBDD-based (Ordered BDD) symbolic model checking algorithm with breath-first-search state traversal [25]. SMV checks for safety, liveness, fairness and deadlock freedom. A system is specified in the SMV input language. The input language allows the description of asynchronous as well as synchronous state systems, i.e. network of asynchronous, non-deterministic processes or Mealy machines. The language is restricted to finite data types. CV (CMU VHDL temporal logic model Checker) is a symbolic model checker for VHDL based on the concepts of SMV [27].

FDR (Failures-Divergence Refinement) is a Standard ML-based refinement checker and model checker for untimed CSP specifications and implementations [28]. FDR failures are sets of events that a process can refuse when being in a particular state. Divergences are states on which a process can perform an infinite sequence of actions. FDR checks for safety, liveness and refinements on the basis of traces. FDR provides 3 main functions: (1) transform specification into normal form, (2) divergence checking of the implementation and (3) model checking of specification and implementation. For checking states against traces, the transition system defined by a CSP specification is transformed into an equivalent normal form. This is mainly done to eliminate non-determinism and to make all actions visible. The model checker checks the validity of implementation states and failures with respect to the specification in normal form. The check is based on a full and explicit expansion of the state space applying breadth-first search.

3.2 Theorem Prover

Theorem proving is the process of finding a proof for a given set of axioms and inference rules. There are two approaches to theorem proving which have reached wide acceptance and can be denoted as the classical approaches: *HOL* and the *Boyer-Moore Theorem Prover*. An alternative approach implements the State Delta Verification System (SDVS) of the Aerospace Corporation. SDVS is based on state delta logic, which is a variant of temporal logic. SDVS provides theorem proving for descriptions written in subsets of ISPS, Ada and VHDL [29].

In the following we first introduce the classical HOL and Boyer-Moore approach. Thereafter, we investigate the Prototype Verification System (PVS) and the Stanford Temporal Prover (STeP).

HOL is a model-based interactive proof assistant for high order logic [30]. The HOL approach is based on Milner's LCF (Logic for Computable Functions). In HOL, theorem proving is mainly applied for checking design refinements (implementation vs. specification). The specification is given as a set of axioms. The implementation represents the theorems that have to be interactively deduced from the axioms applying a set of specified inference rules. The logic of HOL is based on set theory and typed predicate logic. It extends first-order logic by equality, conditions and permits high-order functions (λ-expressions). Variables can range over functions, predicates and quantification over arbitrary types. All terms in high-order logic have a type. Basic HOL has the atomic types, boolean and natural numbers, as well as compound types, function types and a number of constants (e.g. implication and polymorphic equality). The constants come with a set of axioms and theorems. Temporal properties are defined as predicates on the execution. Predicates are given as typed sets. A small set of predefined inference rules support the deduction. Inference rules in HOL refer to functions, which return theorems if proper arguments are given. The system allows forward proofs as well as backward proofs based on the sequent calculus. In contrast to classical logics, theories are dynamic extendible objects in HOL.

BMTP (Boyer-Moore Theorem Prover) [30] is based on the principles of mathematic induction. Nqthm and ACL2 are both successors of BMTP [31,32]. In contrast to HOL, Nqthm provides a different style of theorem proving. Nqthm is a user-guided automatic deduction tool for checking properties of a specified system. The system has to be specified in terms of inductively defined total functions. Nqthm is based on propositional logic with equality. The basic theory includes axioms defining boolean constants for true and false values, equality, if-then-else functions and Boolean arithmetic operations (conjunction, disjunction, negation, implication and equivalence). The theorem prover takes a term in propositional logic as input and reduces the term by the means of mathematic induction automatically generating induction schemes. Many heuristics and decision procedures are implemented as part of the transformation mechanism.

PVS (Prototype Verification System) from SRI (Stanford Research Institute) International Computer Science Laboratory is a theorem prover written in Standard ML [33]. The PVS specification language is based on high-order predicate logic. PVS specifications are divided into (parameterized) theories with assumptions, definitions, axioms and theorems. PVS expressions cover usual arithmetic and logical operators, function application, lambda abstraction and quantifiers. Constraints, such as the type of odd numbers, are introduced by the means of predicate subtypes and dependent types. The PVS theorem prover provides a collection of powerful primitive inference procedures that are interactively applied under user guidance based on a sequent calculus. The primitive inferences include propositional and quantifier rules, induction, rewriting and decision procedures. PVS integrates experimental integration of CTL model checking. Some inductive steps can be automatically discharged by the model checker.

STeP (Stanford Temporal Prover) integrates a model checker and an automatic deductive theorem prover [34]. The input is given as a set of temporal formulae and a transition system that is generated from a description in a reactive system specification language (SPL) or a description of a VHDL subset. The formulae and the transition system are the input for both, a model checker and an automatic theorem prover. The prover covers techniques based on first-order theorem proving, term rewriting, decision procedures and invariants. Interactive proofs are constructed with verification diagrams. STeP is for the verification of parameterized (N-component) circuit designs and parameterized (N-process) programs, including programs with infinite data domains.

4. SEMI-FORMAL NOTATIONS

In contrast to formal languages, semi-formal notations are notations that provide a set of symbols to represent specific roles in the description of a system, but have a loosely defined semantics. The use of a syntactically consistent notation generally brings a more formal feel to descriptions of systems than a natural language description would. This can be misleading as the lack of a precise semantics leaves the description open to different interpretations.

4.1 The UML

The Unified Modelling Language [35] emerged as a standardisation of the leading object-oriented analysis and design methods that were competing for favour in the late 1980s and early 1990s. Responsibility for the standardisation has been taken over by an independent consortium, the Object Management Group (OMG). Several software tool manufacturers market tools to support the use of the UML.

The UML is a notation for use in modelling object-oriented designs and consists of 13 diagrams. The most important one for system specification and modelling are:

- *Use case diagrams* are a means of organizing requirements descriptions into event sequence scenarios. A scenario is triggered by an actor (an external object such as a person interacting with the system) and parts of the system's responsive actions are then packaged and represented by named symbols. The meaning of a particular symbol is defined textually, usually in natural language.
- *Class diagrams* are used to model the static structure of a problem or system. Entity types are represented by classes and the relationships between them are shown as associations and generalizations. Classes represent sets of like instances and are given attributes that represent state variables and values associated with each instance of the class. Classes also have operations that define how an instance's attributes and associations alter in response to events.
- *Collaboration diagrams* and *sequence diagrams* are similar to each other. They both show dynamic behaviour as objects (of the classes introduced in the class diagram) interacting, by passing messages or calling each other's operations, to perform a particular behaviour or task scenario. Sequence diagrams show the interaction as a time ordered sequence of messages passed between objects. Collaboration Diagrams show the same sequence of messages but overlaid on a network of connected objects rather than a time sequence. Note here, that UML 2.0 introduce several advanced features to sequence diagrams, like control structures by combined fragments, which are not available in collaboration diagrams.
- *State diagrams*[2] and *activity diagrams* – Statechart/Activity models, constructed and viewed via state diagrams and/or activity diagrams,

[2] In UML 2.0, state diagrams (and statecharts) are called state machines.

show behaviour in terms of a set of states and transitions between them. Each transition can be annotated with the event that causes it to occur, any guards, which must be true before it can occur and actions that are performed when it occurs. Activity diagrams are a development from state diagrams that also allow "forks" to activate more than one state simultaneously and synchronizations that require more than one state to be active before a transition can occur (when drawing activity diagrams, states are called activities). Statechart/Activity models can be used at several levels. For example, they can be attached to the logical model, to use cases or to classes. Note here, that UML 2.0 introduces a major revision to the unification of state, activity and sequence diagrams also covering Petri-Nets semantics.

4.2 Integrating Formal and Semi-Formal Notations

Semi-Formal Notations such as UML are gaining widespread popularity in industry but lack precision for describing detailed behaviour unambiguously. Conversely, formal notations have not gained widespread use in industry despite their recognized benefits. An integration of semi formal and formal notations may address the deficiencies of the semi formal notations while making the formal notation more approachable. Craigen, Gerhart and Ralston [36] found that better integration of formal methods with existing software assurance techniques and design processes was commonly seen as a major goal; they concluded that *successful integration is important to the long term success of formal methods*. Fraser, Kumar and Vaishnavi [37] discuss some of the reasons why this may be true and go on to describe a framework for classifying current formal specification processes according to the degree of transitional semiformal stages. The categories are direct (no transitional stages), sequential transitional (transitional stages developed prior to the formal specification) and parallel successive refinement (formal specification derived in parallel with semiformal specification through iterative process). Paige [38] analyses the composition of compatible notations and derives a meta-method for formal method (and semi-formal method) integration. Jackson [39] has developed a formal notation, Alloy and associated tool Alcoa. The Alloy notation has a partial graphical equivalent notation in which state can be expressed. This can then be converted into the textual version of the notation where operations can be added and analyses performed. Without tools to investigate the implications of different structures however, the graphical format is limited to illustration of structure. Several research groups have developed integration between graphical object-oriented notations, including the UML and formal notations such as B [13] and Z [40]. The precise UML group[3] is a collaborative effort to precisely define UML semantics via formalization. The object constraint language, OCL [41] is a formal notation that is part of the UML. It can be used to attach formal constraint statements to elements of UML models to constrain their values. For example, the behaviour of an operation can be precisely defined by attaching OCL statements for the pre- and post-conditions of the operation.

5. APPLYING FORMAL METHODS IN SYSTEM DESIGN

The approaches reported in literature cover most of system design phases. Tools from industry and academia are also available for supporting the design of complex systems. They fall into the following categories:

- *Tools that support validation through simulation* Indicative examples are the SDL suite of Telelogic's Tau [42], which supports simulation of SDL system models and iLogix's Statemate MAGNUM [43] that allows simulation of Statechart models using specific test plans, while co-simulation among system models described in different specification languages is also available. A co-simulation approach using instruction set simulators, extended with the required dummy hardware models has been proposed for co-simulation of mixed hardware/software systems. CoWare [44] and Seamless [45] are typical co-design tools used in industry for hardware/software co-simulation and co-verification. The system description is given in VHDL or Verilog for hardware and C for software. Both allow co-simulation between hardware and software at the same abstraction level.
- *Tools supporting validation via test generation* AGATHA [46] belongs in this category and supports validation of system specifications. The system models can be described either in UML or in SDL or in Statecharts. Based on the system specifications, AGATHA automatically generates symbolic test cases, which are used as input to the initial system model in order to detect design weaknesses like deadlocks. Telelogic's TTCN suite [47] offers TTCN[4] testing of SDL system models and simulation of test suits (set of TTCN tests) referring to a specific test case. Esterel studio [48] allows specification and validation of complex systems using a validation

[3] Available at: *http://www.cs.york.ac.uk/puml/maindetails.html.*

[4] Tree and Tabular Combined Notation, or Testing and Test Control Notation in TTCN-3, is an ETSI standardized language for describing "black-box" tests for reactive systems such as communication protocols and services.

engine for checking the consistency of the implementation with the initial specifications, while it supports automatic test coverage generation for the system under design.

- *Tools that support formal verification* TNI-Valiosys consultant applies a formal verification technology, called Linear Programming Validation (LPV), at customers' models (mainly described in SDL) in order to validate and debug the behaviour of the SDL architectural model [49]. The purpose of this approach is to guarantee that the SDL model behaviour respects the specification requirements of the system. A similar approach has been adopted by imPROVE-HDL [49], a high performance model checker dedicated to hardware block validation, for white box and black box verification. The checker accepts as input a VHDL/Verilog description of the component and a description of the block's properties. The result is an indication of the block's compliance with the required properties, or a counter example to highlight problematic behaviours. For hardware implementations, SOLIDIFY [50] allows exhaustive functional verification of the system under design at the RTL level, aiming at reducing the test vectors needed for system level verification. The initial models of the system are described in a HDL like Verilog or VHDL.

The common denominator of most design practices is that despite the plethora of methods and tools supporting designers during various phases, most of them rely on simulation for system validation and verification. Simulation at early design stages and hardware/software co-simulation at late design stages are commonly used practices for validating and verifying the system under design.

In order to deal with the increasing verification complexity, there are attempts to apply formal methods for verifying system properties during every design phase. For example, B method/language for designing complex systems has been applied in [51,52], while B has also been applied for the design of hardware components. In [53], Event B is used to specify and refine a circuit, while the approaches in [54,55] model circuits close to an implementation level. The first one requires that the model uses basic logic gates which are modelled in terms of B machines; the second one allows higher data-types like integers in their models, while it introduces new structuring mechanisms into B which mirror those of VHDL closely.

The next book chapters introduce and describe thoroughly the *PUSSEE (Paradigm Unifying System Specification Environments for proven Electronic design) method*, which applies the main concepts for integrating formal and semi formal methods for embedded system design. In that context, the combined use of UML and B is introduced while a set of supporting tools is also described.

REFERENCES

1. U. Glässer, *Systems Level Specification and Modeling of Reactive Systems: Concepts, Methods and Tools*, Proceedings of EUROCAST 95, Springer Verlag, 1995.
2. IEEE, *Software Engineering Standards,* The Institute of Electrical and Electronics Engineers, 1987.
3. G. Winskel, *The Formal Semantics of Programming Languages*, The MIT Press, 1993.
4. C.A.R. Hoare, *An Axiomatic Basis for Computer Programming*, Communications of the ACM, 12(10), 1969.
5. J.E. Stoy, *Denotational Semantics: The Stoy-Strachey Approach to Programming Language Theory*, MIT Press, 1977.
6. H.P. Barendregt, *The Lambda Calculus, Its Syntax and Semantics*, North Holland, 1981.
7. I. Sommerville, Editor, *Software Engineering*, 4th Edition, Addison Wesley, 1992.
8. C.J. Hogger, *Essentials of Logic Programming*, Clarendon Press, 1990.
9. M. A. Orgun and W. Ma, *An Overview of Temporal and Modal Logic Programming*, Proceedings of First International Conference of Temporal Logic (ICTL 94), Springer Verlag, 1994.
10. Z. Manna and A. Pnueli, *The Temporal Logic of Reactive and Concurrent Systems*, Springer Verlag, 1992.
11. J. Bowen, *Specification and Documentation using Z: A Case Study Approach*, International Thomson Computer Press, 1996.
12. C.B. Jones, *Systematic Software Development using VDM*, Prentice Hall International, 1990.
13. J-R Abrial, *The B Book: Assigning Programs to Meanings*, Cambridge University Press, 1996.
14. Th. Kropf, *Introduction to Formal Hardware Verification*, Springer Verlag, 1998.
15. L. Lamport, *The Temporal Logic of Actions*, ACM Transactions on Programming Languages and Systems, 16(3), May 1994.
16. A. Arnold, *Finite Transition Systems*, Prentice Hall Int., 1994.
17. R. Milner, *Communication and Concurrency*, Prentice Hall, 1989.
18. C.A.R. Hoare, *Communicating Sequential Processes*, Prentice Hall, 1985.
19. G. J. Milne, *Circal and the Representation of Communication, Concurrency and Time*, ACM Transactions on Programming Languages and Systems, 7(2), 1985.
20. The RAISE Language Group, *The RAISE Specification Language*, Prentice Hall, 1992.
21. P.D. Mosses, *Action Semantics*, Number 26 in Cambridge Tracts in Theoretical Computer Science, Cambridge University Press, 1992.
22. Y. Gurevich, *Evolving Algebras 1993: Lipari Guide*, In E. Börger, Editor, Specification and Validation Methods, Oxford University Press, 1994.
23. E. Börger and R. Stärk, *Abstract State Machines - A Method for High-Level System Design and Analysis*, Springer Verlag, 2003.
24. E. M. Clarke, O. Grumberg and D.A. Peled, *Model Checking*, MIT Press, 2000.
25. R. E Bryant, *Binary Decision Diagrams and Beyond Enabling Technologies for Formal Verification*, Proceedings of the 1995 IEEE/ACM Conference on Computer Aided Design, ACM Press, 1995.
26. K.L. McMillan, *The SMV System*, Available at: http://www-2.cs.cmu.edu /~modelcheck/smv.html, 2004.
27. CV, *A Model Checker for VHDL*, Available at: http://www-2.cs.cmu.edu /~cmuvhdl/index.html, 2004.
28. Formal Systems (Europe) Ltd, *FDR2 - User Manual and Tutorial*, Available at: http://www.fsel.com/documentation/fdr2/html/, 2003.

29. I. Filippenko, *Some Examples of Verifying Stage 3 VHDL Hardware Descriptions Using the State Delta Verification System (SDVS)*, Technical Report ATR-93(3738)-3, The Aerospace Corporation, 1993.
30. R. S. Boyer and J. S. Moore, *A Computational Logic*, Academic Press Inc, 1979.
31. R.S. Boyer and J. S. Moore, *A Computational Logic Handbook*, Academic Press Inc., 1988.
32. M. Kaufmann, P. Manolios, J.S. Moore, *How to Use ACL2*, Kluwer Academic Publishers, 2000.
33. J. Rushby, *The PVS Verification System*, Available at: http://www.csl.sri.com/sri-csl-pvs.html, 1995.
34. Z. Manna, A. Anuchitanukul, N. Borner, A. Browne, E. Chang, M. Colon, L. de Alfaro, H. Devarajan, H. Sipma and T. Uribe, *STeP the Stanford Temporal Prover*, Technical Report STAN-CS-TR-94-1518, Stanford University, 1994.
35. J. Rambaugh, I. Jacobson and G. Booch, *The Unified Modeling Language Reference Manual*, Addison-Wesley, 1998.
36. D. Craigen, S. Gerhart and E. Ralston, *Formal Methods Reality Check: Industrial Usage*, IEEE Transactions on Software Engineering, vol. 21, No.2, 1995.
37. M. D. Fraser, K. Kumar and V. K. Vaishnavi, *Strategies for Incorporating Formal Specifications in Software Development*, Communications of the ACM, vol. 37, No. 10, 1994.
38. R. Paige, *Formal Method Integration via Heterogeneous Notations*, Ph.D Thesis, University of Toronto, 1997.
39. D. Jackson, *Alloy: A Lightweight Object Modelling Notation*, Technical Report 797, MIT Lab for Computer Science, 2000.
40. I. Houston and S. King, *CICS Project Report: Experiences and Results from the Use of Z in IBM*, Proceedings of The 4th International Symposium of VDM Europe. Vol. 1, Springer-Verlag, 1991.
41. J. Warmer and A. Kleppe, *The Object Constraint Language: Precise Modeling with UML*, Addison-Wesley, 1999.
42. SDL Suite, Available at: http://www.telelogic.com/products/tau/sdl/index.cfm, 2004.
43. Statemate, Available at: http://www.ilogix.com/products/magnum/index.cfm, 2003.
44. CoWare Inc, Available at: http://CoWare N2C Method Manual. Version 3.1, 2001.
45. Seamless, Available at: http://www.mentor.com/seamless/, 2003.
46. D. Lugato et al, *Validation and Automatic Test Generation on UML Models: The AGATHA Approach*, Electronics Notes in Theoretical Computer Science 66 No.2, 2002.
47. TTCN Suite, Available at: http://www.telelogic.com/products/tau/ttcn/index.cfm, 2003.
48. Esterel Studio. Available at: http://www.esterel-technologies.com/v3/, 2003.
49. Valiosys, Available at: http://www.tni-valiosys.com/, 2003.
50. Solidify, Available at: http://www.saros.co.uk/, 2003.
51. J. Draper et al, *Evaluating the B method on an avionics example*, Proceedings of Data Systems in Aerospace (DASIA) Conference, 1996.
52. C. Snook, L. Tsiopoulos and M. Walden, *A Case Study in Requirement Analysis of Control Systems using UML and B*, Proceedings of International Workshop on Refinement of Critical Systems, Methods, Tools and Developments, 2003.
53. J.-R. Abrial, *Event Driven Electronic Circuit Construction*, Available at: http://www.atelierb.societe.com/ressources/articles/cir.pdf
54. S. Hallestrede, *Parallel Hardware Design in B*, Proceedings of 3rd International Conference of B and Z Users, Lecture Notes in Computer Science, vol. 2651, Springer-Verlag, 2003.
55. W. Ifill et al, *The Use of B to Specify, Design and Verify Hardware*, In High Integrity Software, Kluwer Academic Publishers, 2001.

Chapter 2

FORMALLY UNIFIED SYSTEM SPECIFICATION ENVIRONMENT WITH UML, B AND SYSTEMC

Klaus Kronlöf, Ian Oliver
Nokia Research Center

Abstract: A methodology for introducing formal proof to embedded systems development. It is based on the formal semantics of B, uses UML notation as the primary design language, provides a formally provable decomposition techique, applies model checking for the verification of temporal properties and supports hardware synthesis in addition to software implementation.

Key words: System design, formal methods, B, UML, model checking, hardware synthesis

1. GENERAL PRINCIPLES

Initially when developing the methodology we had quite definite idea of what the ideal process of system design looks like and we stated our objectives accordingly in terms of this fixed process. Later on when working on various case studies we learned that development processes in different domains vary a lot. In the end we restated our objectives such that instead of one fixed process we provide building blocks for defining domain specific processes.

In this document the concepts of model-based development and OMG's (Model Driven Architecture) MDA have been used to define the methodology. We decided to refer to MDA in this final description despite it was not explicitly present in our original plans, because it actually fits quite nicely to the context and it is well known to the software community.

The general idea in the method is that the formal semantics is based on the B method while UML is used for user interface in order to lower the

J. Mermet (ed.), UML-B Specification for Proven Embedded System Design, 21–35.
© 2004 *Kluwer Academic Publishers. Printed in the Netherlands.*

industrial deployment barrier. This basic core is complemented by more domain-specific languages for analysis (model-checking) and HW/SW implementation,

2. METHODOLOGICAL INNOVATIONS

The method is based on four major innovations:

- A UML-B profile that connects semantically the underlying meta-models of UML and B
- A decomposition technique for provably correct logical architecture
- A method for formal verification of (logical) timing properties using a model checking tool (RAVEN)
- A method for hardware design with B, including BHDL subset supporting hardware synthesis

3. MDA MAPPINGS

The basic idea behind MDA as described in the OMG's documentation is that models are related by mappings. Current versions of the model describing this structure are exceedingly complex (and probably with good reason). Here we present a much simplified version of this structure which concentrates on the relationship between the concept of a model, the mappings and the language in which the model is written in.

We describe the MDA's structure as the MDA Meta-Model. In the diagram below we can see this structure described using the UML class diagram.

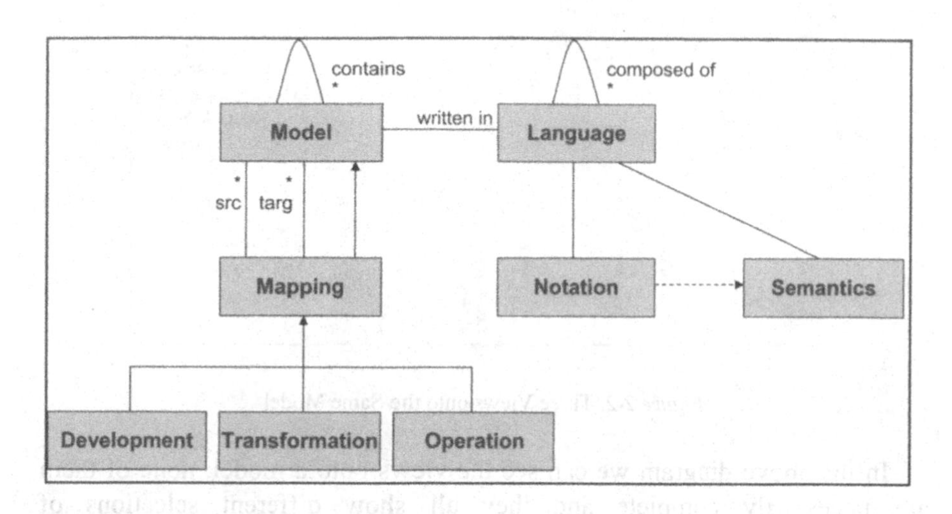

Figure 2-1. MDA Meta-Model

The model can be divided into three parts:

- The Model
- The Mappings
- The Languages

Each will now be discussed in more detail.

3.1 Model

The concept of a model comes from the UML's idea of a model which in turn is based upon the ideas from software engineering as a whole.

A model is a collection of modeling elements (which themselves maybe models). A model is never (unless very simple) viewed in its entirety. Normally one only views part of a model in which one is interested, for example, a particular set of classes and their relationships. Each view onto the model shows only a partial, incomplete view of the model. Only if the view is denoted to be complete can we be sure that all information about that particular aspect of the model is shown in the view.

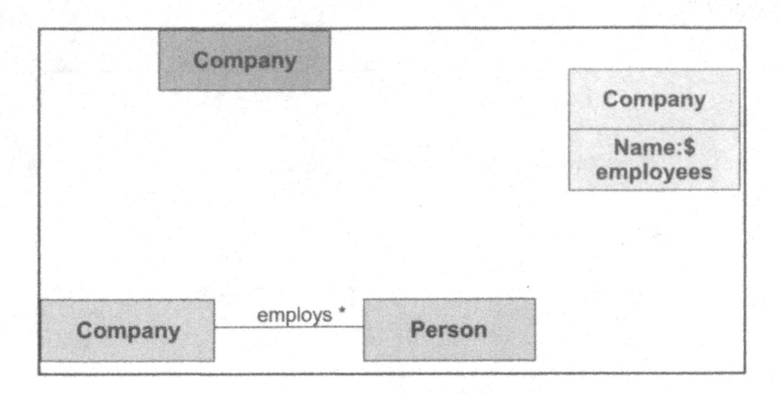

Figure 2-2. Three Views onto the Same Model

In the above diagram we can see the views onto a model, none of them are necessarily complete and they all show different selections of information. The underlying model may also contain information about the states of particular classes, other classes, subtypes, attributes, workflows, contract etc. To obtain this information one must consult the model using other views.

Models of course may contain relationships to other models - these models could describe different versions of the models at differing development stages, different parts being worked upon by different groups of developers and so on.

In the method presented in this chapter the aim has been to define a UML profile, called the UML-B, the includes a complete set of views to the model of the system under design. The current UML-B profile does not achieve this fully, for example it still lacks the expression of real-time properties.

3.2 Mappings

One of the core concepts in the MDA is that of a mapping. A mapping is some algorithmic method of moving between models. Inside the concept of a mapping is the idea of a platform which is the target environment and also the concepts of PIM and PSM - platform independent model and platform specific model respectively. A mapping is a device that takes a platform-independent model and maps it to a platform specific model.

The MDA at present does not describe any form of taxonomy or ontology of these mappings or how one is constructed. As a simple taxonomy we define here three types of mappings: Development mappings, transformation mappings and operation mappings. The first two are most relevant to the method presented here. The figure below depicts the mappings of the method.

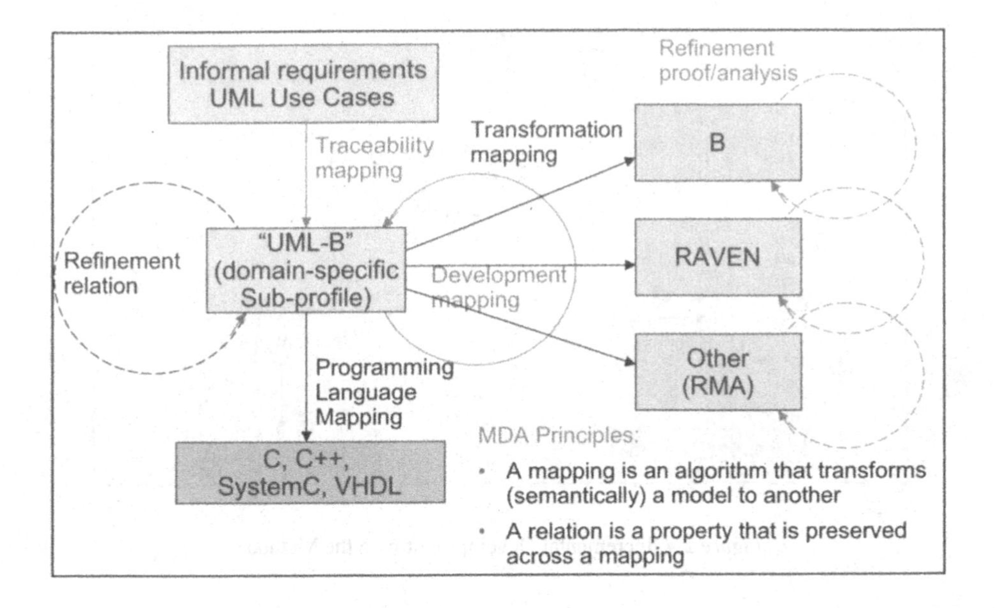

Figure 2-3. MDA Mappings of the Method

3.2.1 Development Mappings

We define a development mapping as one that takes a platform independent model and generates a platform specific model. This is the form that it envisaged by the MDA. In the method presented here we regard incremental refinement as a series of development mappings each of which is a PIM-PSM mapping in relation of the relevant platform at that level of abstraction. A development mapping is a refinement, if the pair of models satisfies the chosen refinement relation. In the method the refinement relation is defined by the B method.

A consideration of lower lever development mappings is the efficiency of the generated model (read code as being a model). While the advantages of keeping a large semantic gap are obvious - models are implementation language independent - under some circumstances, for example in embedded systems the amount of information in the development mapping to facilitate an efficiently coded mapping implies that the mapping becomes very complex and tied to one particular use under very specific circumstances. This can be avoided by splitting development mappings into a number of more generic discrete steps. Our approach is to leave the choices of size (semantic gap) and number of steps to the designer. The figure below development depicts a typical path.

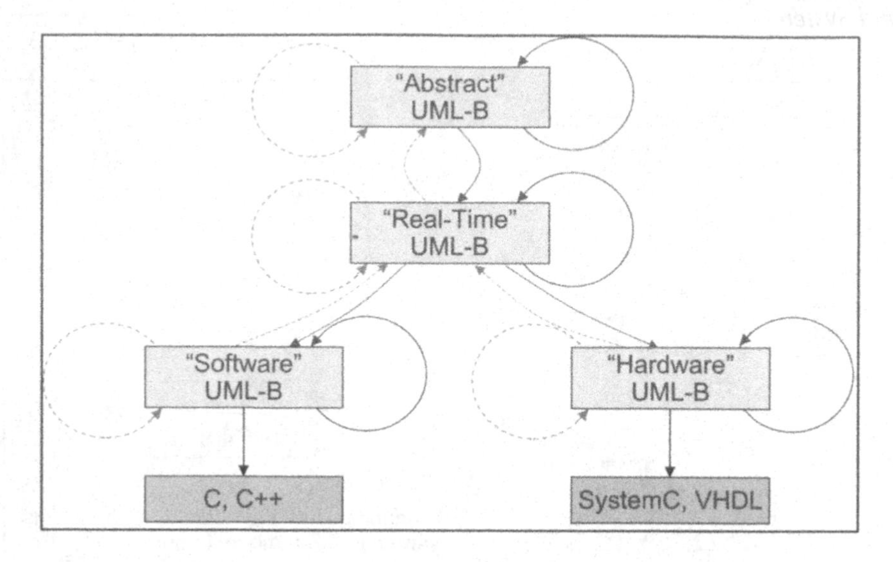

Figure 2-4. Incremental Development with the Method

3.2.2 Transformation Mappings

Development mappings can be considered to be vertical in nature - that is they drive the move from abstract models to concrete models - a transformation mapping on the other hand we consider to be horizontal in nature in that is does not drive the development of the model but transforms it into another representation.

The reasons for this are primarily to facilitate the extraction of certain kinds of information from the model for processing in another form. A good example of this is in real-time system development it is a good idea to understand the schedule-ability characteristics of the model. There exist mappings from UML-RT to Rate Monotonic Analysis (RMA) which is a schedule-ability theory. One can take any UML-RT model if suitably annotated and build a RMA model which only takes into consideration the timing characteristics. From this model we can perform a schedule-ability analysis and decide further development upon the results.

Transformation mappings are not necessarily reverse-able – that is the idea of "round-trip" style engineering between two different models written in two different languages is not always practical or desirable. In most circumstances the transformed model will be used as the basis for further development decisions rather than an explicit modeling medium.

The change of language between the source and target models here can be quite extreme but the models still do remain at the same level of abstraction - there is no concept of platform independent and platform specific here.

In our method we have defined a transformation mappings from the UML-B profile to procedural and event B models. We have also studied the transformation mapping to a model that supports model-checking of real-time properties.

3.2.3 Operation Mappings

This kind of mapping is not a mapping in that it drives the model development or transformation as described previously. These mappings correspond to operations that can be performed upon a model such as model intersection, model union, equality, isomorphism and so on. There is research progressing in this area which is known as model calculi.

Common examples of these kinds of mapping can be found in source code management systems such as CVS where the concepts of code equality and code union exist.

These kinds of mappings will probably not appear in tools as the development or transformation mappings in that they generate explicit models but rather be implemented as queries upon particular sets of models. In more formal development an example of this would to be to ask the question of a pair of models related by a development mapping "does the platform specific model refine the platform independent model?" or "does the platform specific model correspond behaviorally to that specified in the platform specific model?"

3.3 Language

Every model is written in some kind of language. In the case of MDA it is suggested by the OMG that the language used is of the OMG family of languages, that is: UML or one of its profiles (UML-RT, SPEM etc). The situation is complicated by the fact that the OMG only have a small set of general-purpose languages that are too generic in their purest form for many purposes.

Any language may be used to write a model, for example the formal specification languages such as Z, B and VDM are may be used to specify critical properties, SDL, LOTOS, Petri-Nets and other process calculi languages are aimed at behavioral properties. Finally of course there is the extensive set of implementation languages such as C++, Java, C, Eiffel, Smalltalk, Prolog, LISP, Fortran, COBOL and so on.

Languages are not as simple as they appear - they are not just a collection of symbols (syntax) but contain meaning (semantics) that define how the syntax is constructed and used.

Until the UML most graphical modeling languages tightly bound the syntax and semantics. The UML takes great care to separate the two concepts. This is both the UML's advantage and one of its biggest problems. It is necessary to utilise the UML properly that the developer understands that particular sets of semantics should be combined at differing levels of abstraction in order to give the models the correct meaning for the development work in hand.

We now discuss in more detail the concepts of Notation and Semantics which are critical to understanding the concept of language.

3.3.1 Notation

This gives the language its user interface and may be graphical or textual in nature. In the case of the UML it is primarily graphical but also contains textual components in the form of OCL and notes.

An important theoretical result is that graphical notations are not as express-able as textual notations. The details of the mathematics behind this are exceptionally complex but this is the reason why graphical languages need to be annotated with textual components. The point at which one needs to move to textual specification from a graphical one depends upon the language and the express-ability of the language. This express-ability varies depending upon the semantics. The one important fact here is that one can not just use graphical languages alone to specify systems unless you wish the specification not to contain enough details to be usable. Graphical languages which have high degrees of express-ability are highly application specific and can not be used outside their particular domain - even then they rely upon textual annotations.

In our method the primary notation is defined by the UML-B profile and it is a combination of UML graphics and textual components derived from B.

3.3.2 Semantics

Semantics is the structures that give meaning to the language. Each element in the notation is allied with some concept in the semantics. For example the UML class notation (box with some text inside) is allied with the object oriented class concept.

While considered the biggest problem with the UML, this free semantics is an area where the UML shows its true power. The developer can choose which particular semantics to use for which UML modelling element. UML state diagrams provide a convenient example here.

In the B method the semantics of a UML-B model is defined by the transformation mapping to B.

4. UML-B

A distinguishing feature of our method is the way UML is used in a formally rigorous fashion. We have defined a UML profile, called UML-B, for this purpose. UML-B is a UML 1.4 profile, that:

- restricts the UML notation to a subset and imposes syntactical rules onto it (to overcome the problems due to the fact that B is not object oriented)
- uses the extension mechanisms of UML (e.g. stereotyping) to denote the mapping between UML elements and B concepts
- defines a particular semantics for the models based on B
- The UML-B profile is developed in a way that:
- supports both procedural and event B
- uses graphical notations of UML where appropriate (class diagrams, state charts)
- embeds textual B for purposes where native UML lacks features (set enumerations, invariants, pre-conditions/guards)
- provides alternative class-machine mapping options to handle object oriented-ness problems in different situations

We have developed a prototype tool, called U2B, that implements the mapping from the UML-B profile to pure B. The prototype uses Rational Rose for UML. Figures *2-5*, *2-6* and *2-7* give examples using this tools on how different design concepts are represented in UML-B.

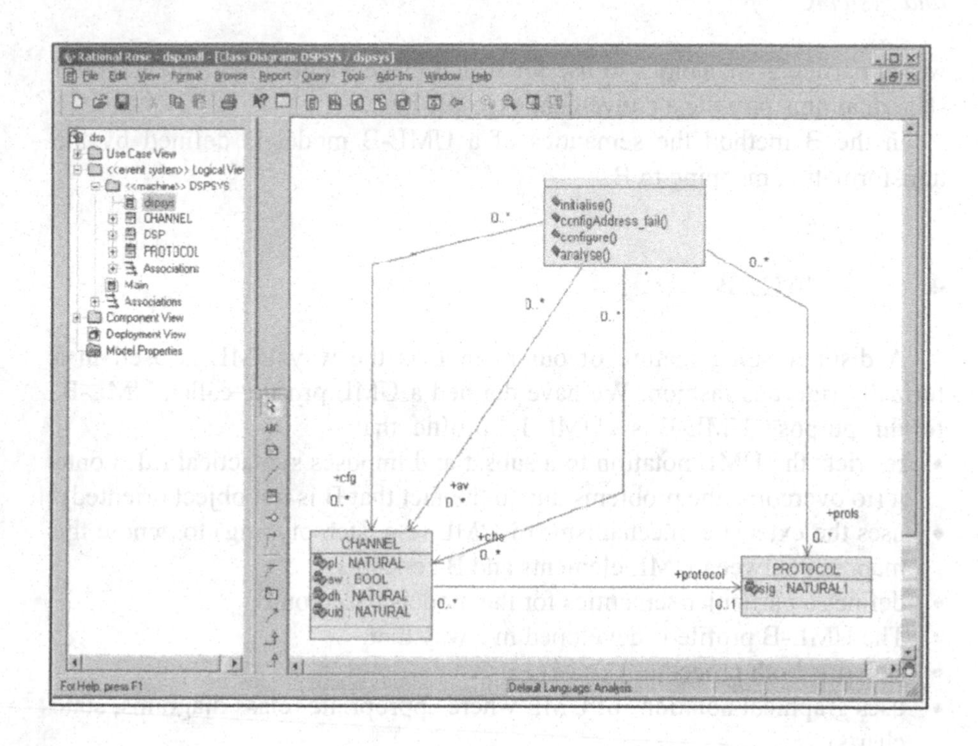

Figure 2-8. Representation of B machines as UML classes in UML-B

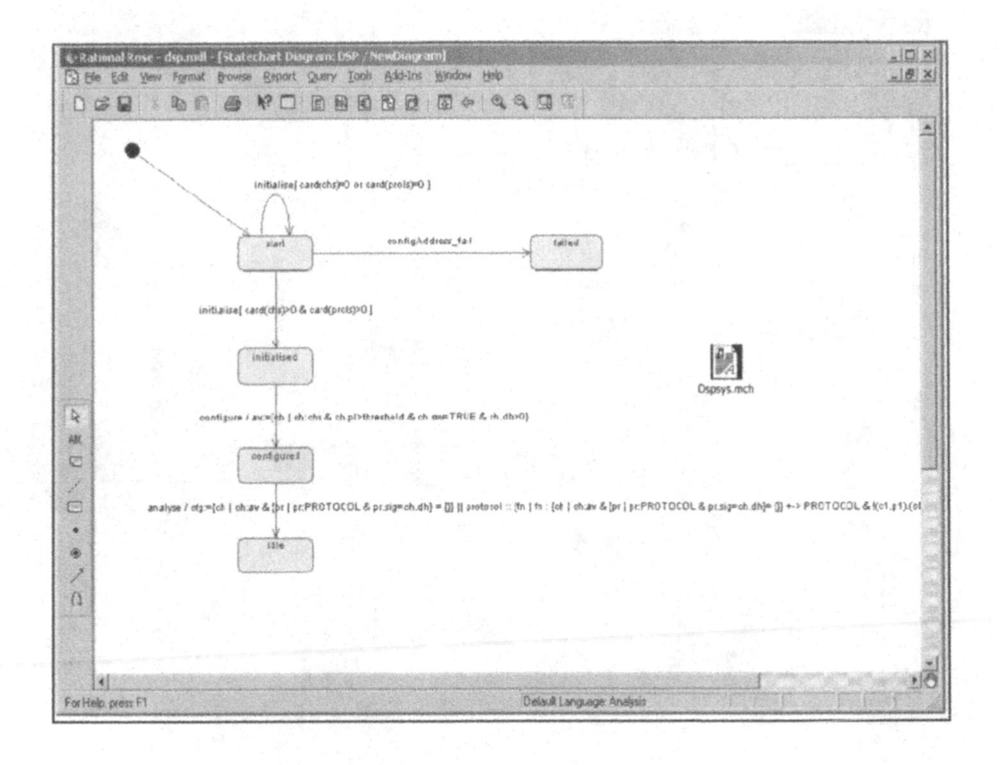

Figure 2-9. Representation of state-oriented behavior in UML-B

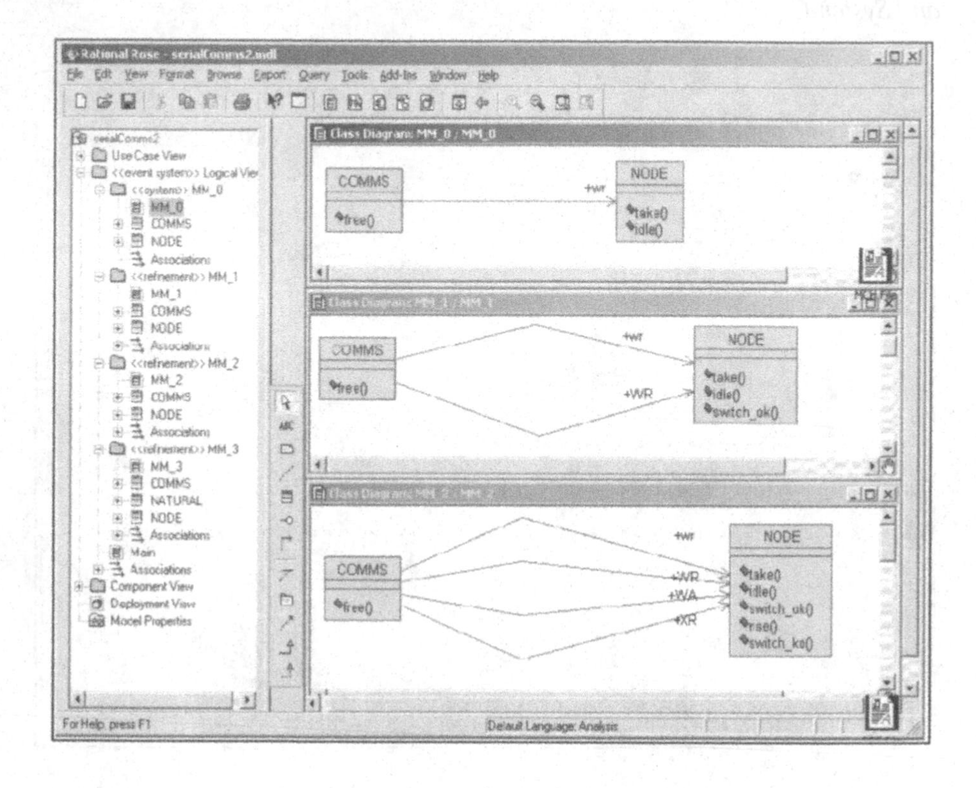

Figure 2-10. Representation of refinement in UML-B

5. DECOMPOSITION

Decomposition is a widely adopted technique in many domains of system engineering. When applied to embedded systems development we can distinguish between logical concerns and physical concerns.

The logical concerns of decomposition are the following:

Breaking monolithic behavior into functional components (logical architecture)

- Management of complexity
- Well defined interfaces
- Minimal communication (overhead)
- Logical correctness

while the physical concerns of decomposition are the following:

- Implementing functionality on existing or developed HW/SW platform (physical architecture)

- Distribution, mapping logical components onto available resources, implementation of communication
- Performance, cost, reliability ...
- Preservation of logical correctness

Our method addresses mainly the logical domain, and the decomposition technique helps especially in complexity management, interface definition and logical correctness.

Figure *2-11* depicts the way decomposition is supported in our method. The basic idea is to provide a technique and tool support that help the designer to decompose the system into subsystems in such a manner that the composite behavior of the decomposed system is provably equivalent to the initial system.

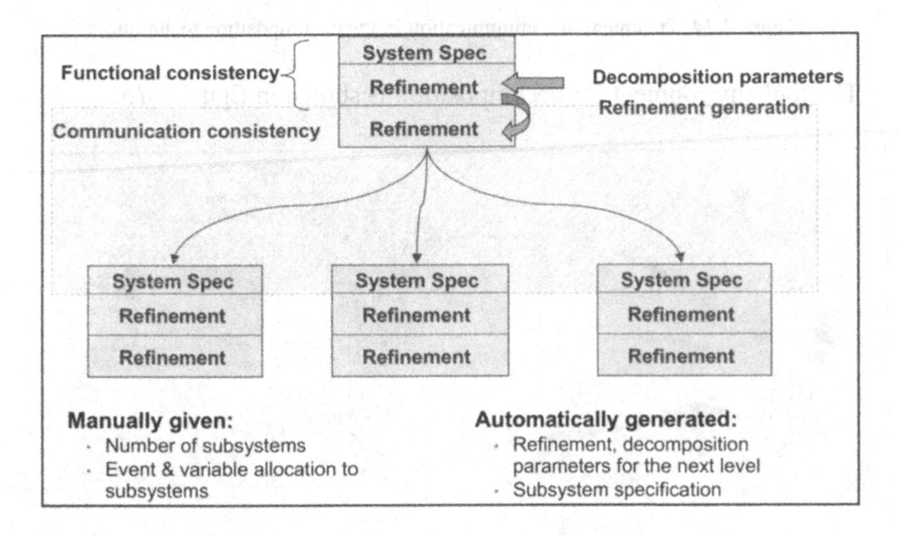

Figure 2-12. Decomposition support

The treatment of communication between subsystems is of special interest in the decomposition technique. Figure *2-13* explains the main ideas. All the variables are distributed in the subsystems in such a manner that each variable is allocated to one and only one subsystem. Built-in protocols are used to take care of references to variables in case it is not allocated to the referring subsystem.

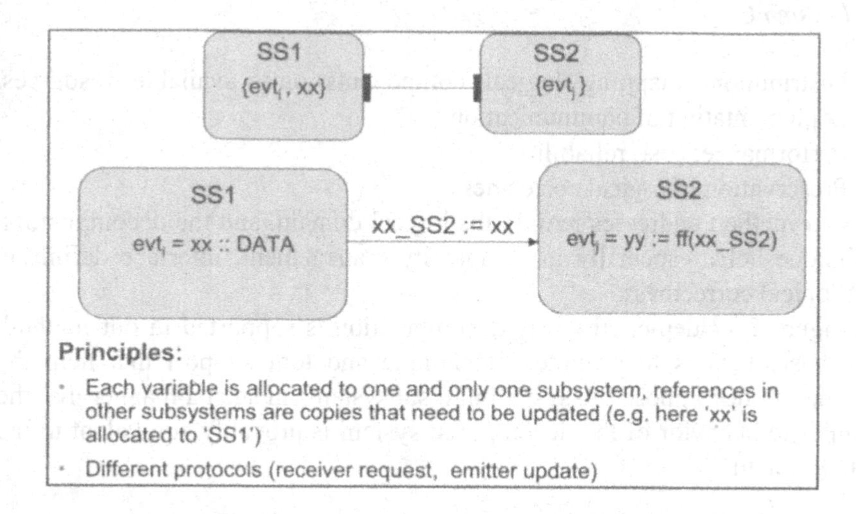

Figure 2-14. Treatment of communication in the decomposition technique

The tool environment for decomposition is shown in figure *2-15*.

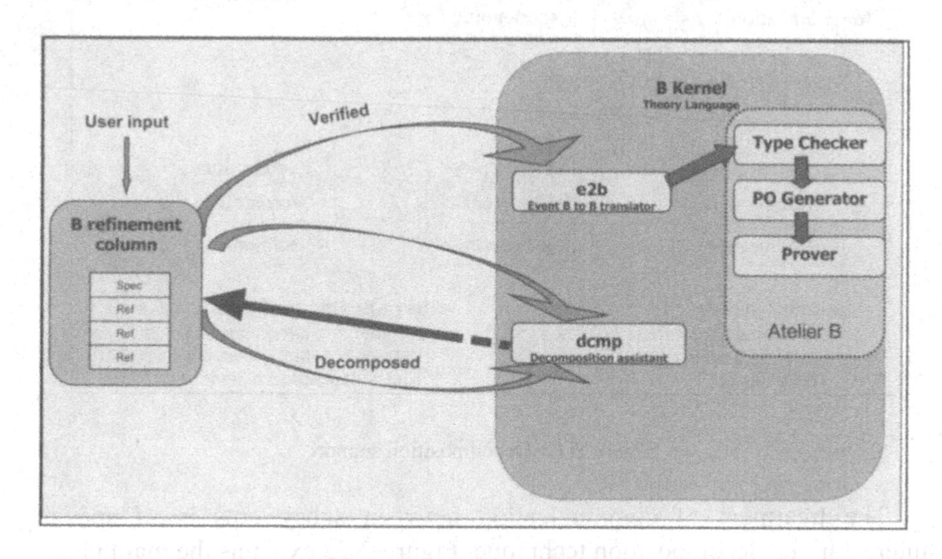

Figure 2-16. Decomposition tools

6. REAL-TIME PROPERTIES

Currently the UML-B profile addresses mainly the functional domain and only very abstract temporal properties based on ordered events can be handled in UML-B. As a step to overcome this limitation we have used a model-checking tool, called RAVEN, to investigate real-time aspects.

RAVEN allows the attachment of time-expressions to state machines and CCTL expressions. These time-expressions denote logical time calculated in computation cycles at so-called cycle-accurate level of abstraction. Using model-checking at this level supports the separation of concerns in real-time analysis. For the verification of logical timing we can use RAVEN and formal model-checking techniques. Then we are left with low level performance verification of physical timing, where we can concentrate on isolated details, since correct cycle level behavior has been formally proven.

We have studied the mapping from RIL (RAVEN Input Language) to B. The proper integration of RAVEN our method would require an extension of UML-B that can be directly mapped to RIL. This has been left as future work. We have also considered replacing RAVEN (which is an academic tool using tool specified languages) with commercial model checker based on the standard PSL notation.

Some readers may know that OMG has defined a standard UML 1.4 profile for expressing real-time properties. It is called Schedule-ability, Performance, and Time Specification (SPT). It includes:

- standard definition of time
- quality of service attributes
- concurrency definitions
- active/passive classes
- signals & port

It is conceivable to extend UML-B with (chosen or all) elements of SPT so that the mapping to RIL can be supported.

7. HARDWARE SYNTHESIS

For hardware implementation purposes we use a subset of the B-language, called BHDL®, for design description. BHDL® consists of:

- a programming notation, in particular the predicate-transformer-based generalised substitutions of B
- structural notations, facilitating modular design and structured proof

BHDL® is designed to support all the abstraction levels that the hardware engineering community is using. Different flavors of BHDL® can be automatically translated by the BHDL-translator into appropriate hardware description languages (HandelC, SystemC, VHDL, Verilog) through a common intermediary format. The synthesize-ability of the generated code has been checked on commercial tools. Being a subsets of the B-language, BHDL® benefits from tool-support for refinement and proof.

Chapter 3

EMBEDDED SYSTEM DESIGN USING THE PUSSEE METHOD

An overview

Nikolaos S. Voros[1]
Colin Snook[2]
Stefan Hallerstede[3]
Thierry Lecomte[4]

[1] INTRACOM S.A., Patra, Greece
[2] University of Southampton, Southampton, United Kingdom
[4] KeesDA S.A., Grenoble, France
[5] ClearSy S.A., Aix en Provence, France

Abstract: The approach presented in this book relies on the unification of system specification environments for developing electronic systems that are formally proven to be correct (correct-by-construction systems). The key concept conveyed is the formal proof of system properties, which is carried out at every phase of the co-design cycle. The main idea is to build fully functional system models that are formally proven to be correct, and based on them to produce automatically the hardware and the software parts of the system. The approach presented relies on the combined use of UML and B language.

Keywords: Co-design, refinement, decomposition, translation.

1. THE UML AND THE B-METHOD

UML is a visual modelling notation for object-oriented systems. Translation to a formal notation that has adequate tool support, such as B [1] enables a model to be formally verified and validated. These verification and validation processes are not avai lable in the UML even if annotated constraints are added in the UML constraint language, OCL [2]. However,

J. Mermet (ed.), UML-B Specification for Proven Embedded System Design, 37–51.
© 2004 *Kluwer Academic Publishers. Printed in the Netherlands.*

translation from unrestricted UML models is problematic because B language is not object-oriented and contains write access restrictions between components in order to ensure compositionality of proofs. There are several approaches reported for mapping between UML to B language [3]. The key differences in our approach are that we specialize UML in order to ensure that the resulting B is amenable to verification. To overcome write-access restrictions we provide a translation from many classes (a package) to a single B component. We also provide UML mechanisms to support an event style of modelling and decomposition [4].

The use of B language for designing complex systems has already been reported in literature [5,6]. As opposed to existing approaches, the co-design method presented in this book proposes the use of B language as part of a unified hardware/software co-design framework that supports formal proof of system properties throughout the various co-design phases.

For the design of hardware system parts, the B language has been applied to circuit design by various authors [7,8]. In [7] Event B is used to specify and refine a circuit but the approach does not provide a translation into a hardware description language. In the context of the work presented in this book we have extended their work in Event B, while we have developed and used a BHDL translator[1] to generate SystemC and VHDL. The approaches in [8,9] model circuits close to an implementation level. The first one requires that the model uses basic logic gates which are modelled in terms of B machines; the second one allows higher data-types like integers in their models, while it introduces new structuring mechanisms into B which mirror those of VHDL closely. The approach presented differs from them in that it does not use an explicit representation for the system clock.

2. PROVEN ELECTRONIC SYSTEM DESIGN USING FORMAL PROOF OF SYSTEM PROPERTIES

The approach introduced is outlined in Figure 3-1. The properties of a system are formally described (and proven) at every design phase. The key aspects of proposed framework are:

System Modelling using a graphical formal notation. For that purpose a specialized profile of UML, called UML-B profile, has been defined in order to allow designers employ UML for defining system models and their properties during early design stages (upper part of Figure 3-1). The

[1] BHDL is a registered trademark of KeesDA S.A., France.

outcome of this process is a set of system models that can be formally proven to be compliant with the initial system specifications.

Formally proven to be correct refinement where the UML-B models already available can be translated automatically to B language. For proving the validity of a model refinement, the B compliant UML models are automatically translated to B where the B language proving mechanisms are employed in order to prove formally the validity of the refinement.

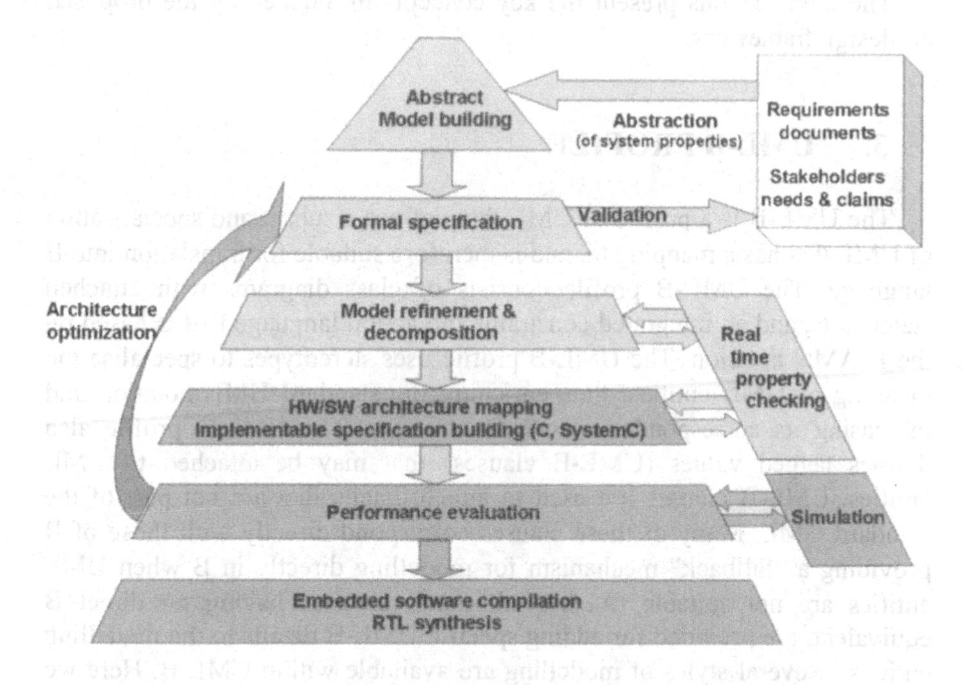

Figure 3-1. Overview of proposed co-design framework (courtesy of KeesDA)

System decomposition into subsystems. Each subsystem can be mapped either to hardware or to software. The initial system is gradually refined and each refinement is proven to be correct. The later is guaranteed through the discharge of all the proof obligations generated (each proof obligation represents a system property that must not violated during refinement). At a specific refinement level, where the system representation is accurate enough, the system is decomposed into subsystems. Each subsystem can be further refined until a fully functional subsystem is reached. The emerging subsystems and the communication between them are described in Event B and they are produced automatically from the decomposition assistant tool developed in the context of the specific approach.

Hardware/software allocation (lower part of Figure 3-1) takes place through direct translations of the emerged subsystems through appropriate translators. The software parts of the system are implemented in C/C^{++}, while the hardware parts of the system are described in VHDL/SystemC. It is important to mention that in both cases, the code produced stems from system models that are formally proven to be correct.

During the last stages of system design, the subsystems produced are simulated together in order to verify their integrated behaviour, while the overall system performance is evaluated. Based on the performance evaluation results, alternative architectures can be explored.

The next sections present the key concepts introduced by the proposed co-design framework.

3. UML-B PROFILE

The UML-B is a profile of UML that defines a subset and specialisation of UML that has a mapping to, and is therefore suitable for translation into B language. The UML-B profile consists of class diagrams with attached statecharts, and an integrated constraint and action language that is based on the B AMN notation. The UML-B profile uses stereotypes to specialise the meaning of UML entities, thus enriching the standard UML notation and increasing its correspondence with B concepts. The UML-B profile also defines tagged values (UML-B clauses) that may be attached to UML entities. UML-B clauses are used to attach details that are not part of the standard UML. Many of these clauses correspond directly with those of B providing a "fallback" mechanism for modelling directly in B when UML entities are not suitable. A few additional clauses, having no direct B equivalent, are provided for adding specific UML-B details to the modelling entities. Several styles of modelling are available within UML-B. Here we use its event systems mode, which corresponds with the Event B modelling paradigm.

UML-B provides a diagrammatic, formal modelling notation. It has a well defined semantics as a direct result of its mapping to B. As with most formal notations there is a strong resistance to the use of B in industrial settings. The popularity of the UML enables UML-B to overcome this resistance. Its familiar diagrammatic notations make specifications accessible to domain experts who may not be familiar with formal notations such as B. UML-B hides B's infrastructure, packages mathematical constraints and action specifications into small sections each being set in the context of its owning UML entity.

4. FORMAL MODEL REFINEMENT

In an event driven approach (like the one presented in this book), refinement performed is only data refinement, as there is no real algorithmic refinement. As described in Figure 3-2, data refinement is based on the introduction of new modelling variables, replacing an existing variable or only complementing the model. The link between the variables of the abstraction and those of the current refinement is located in the invariant and is called *gluing invariant*.

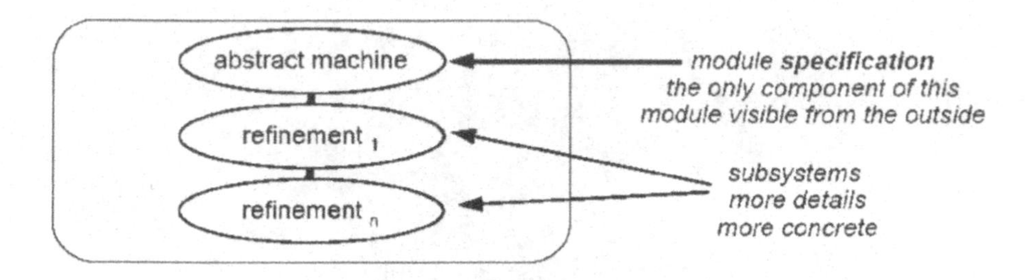

Figure 3-2. Model refinement

Abstraction variables and refinement variables with the same names are considered identical and an implicit gluing invariant is generated ($v_{abstraction} = v_{refinement}$). All new variables must be linked with abstraction variables.

Algorithmic refinement is performed when event based decomposition has lead the designers to identify software/hardware components. Software events for example, can be combined to constitute a single event which represents the specification of one procedure. As such, algorithm specified within this procedure is likely to be refined.

5. BHDL PROFILE

B machines that model hardware, here referred to as BHDL (see Chapter 7), differ in their syntax from B machines that model software. Currently there are BHDL translators to VHDL and SystemC. The VHDL code produced is synthesizable, following the IEEE 1076.6 (draft) standard. The basic data types available in BHDL have been chosen to match with basic SystemC types. The choice of these data types has been made because they are easily portable to other hardware description languages. The substitution language of BHDL is a subset of the substitution language of

B [1]. BHDL machines are expected to be cycle accurate; so while loops have to be encoded using state machines.

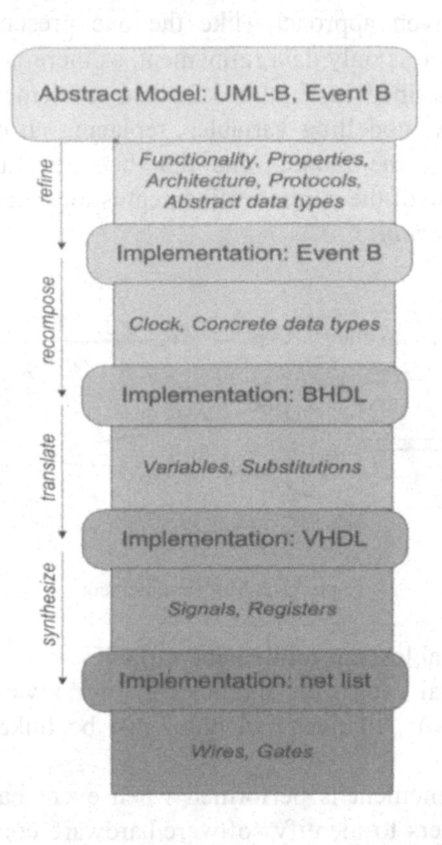

Figure 3-3. Typical BHDL flow

To commence a development on a more abstract level Event B can be used. There is no translation from Event B but a path to BHDL following a simple technique called *event re-composition*. The elements of the BHDL substitution language are interpreted in terms of hardware. Sequential substitution corresponds to causality, simultaneous substitution to concurrency and conditional substitution to multiplexing. Registers are inferred from variables depending on their use. As a rule of thumb, a variable that is read before it is written represents a register. Formally, the translation into hardware description languages is based on the before-after predicate [1], which relates a B substitution to a predicate. Figure 3-3 outlines how the approach is applied to hardware design based on BHDL.

From the abstract model, which is specified in UML-B, we arrive by refinement at an implementable model (still specified in Event B). This model must be cycle-accurate (hw/sw architecture mapping phase in

Figure 3-1). By recomposing events into more complex events we achieve a BHDL model. This event recomposition can be carried out automatically. Once a BHDL model has been produced, this can be translated into VHDL (or SystemC) and subsequently be synthesised. We have successfully synthesised BHDL models using several commercially available VLSI and FPGA synthesis programs. The use of BHDL in this design flow bridges the gap between the often very abstract models produced in Event B and existing synthesis tools. Thus, it delivers within our co-design approach a means to produce correct hardware.

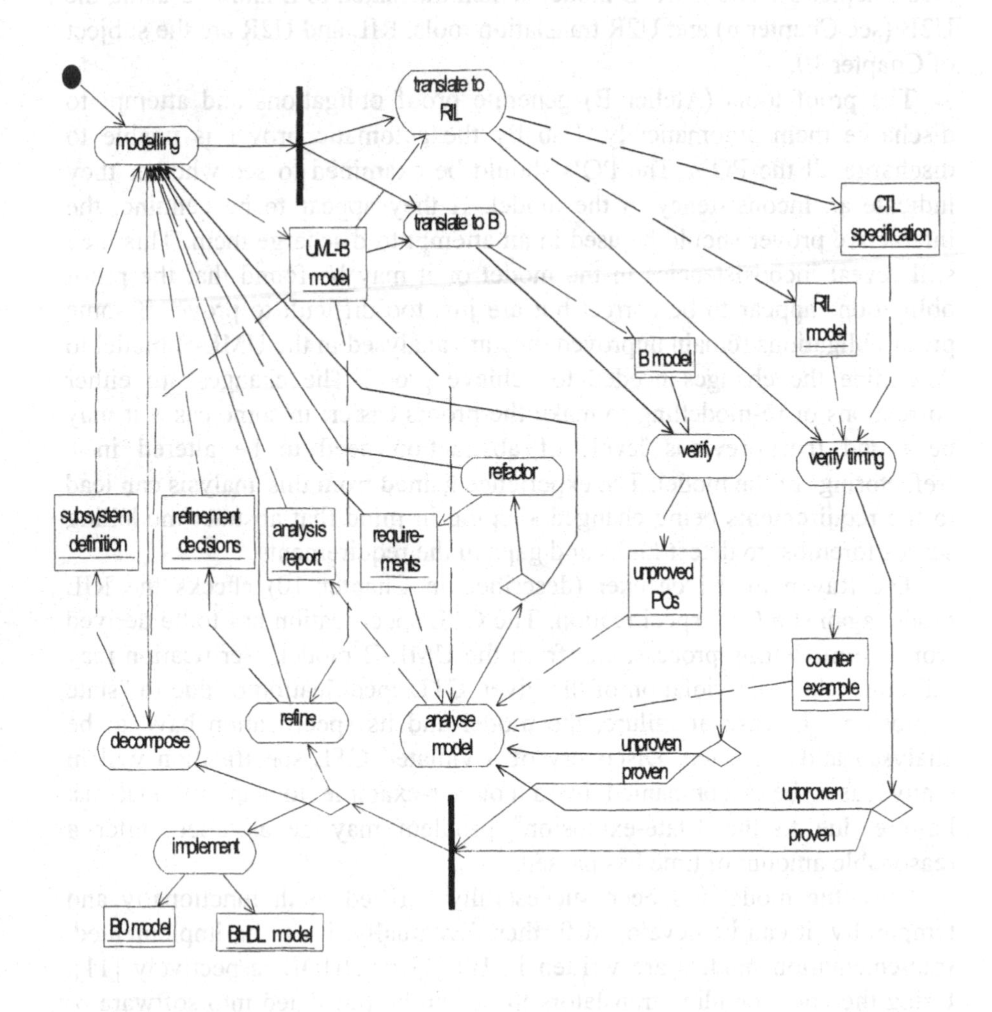

Figure 3-4. Design activities supported by the tool suite

6. THE SUPPORTING TOOLSET

The co-design framework proposed, as outlined in Figure 3-1, is supported by a set of tools that allow system designers to produce systems that are composed of correct-by-construction subsystems. The tools employed in the context of the specific approach cover the all the co-design stages from specification down to implementation.

Figure 3-4 shows this iterative process in more detail, specifically for the notations investigated on the PUSSEE project [10], and illustrates the main information items involved. The semi-formal model is expressed in UML-B (see Chapter 5). The UML-B model is then translated to B and RIL using the U2B (see Chapter 6) and U2R translation tools. RIL and U2R are the subject of Chapter 10.

The proof tools (Atelier B) generate proof obligations and attempt to discharge them automatically. Usually the automatic prover is unable to discharge all the PO's. The PO's should be examined to see whether they indicate an inconsistency in the model. If they appear to be genuine, the interactive prover should be used in an attempt to discharge them. This may still reveal inconsistencies in the model or it may be found that the proof obligations appear to be correct but are just too difficult to prove. If some proof obligations remain unproved they are analysed in the UML-B model to determine the changes needed to achieve proof. The changes are either corrections or re-modelling to make the proofs easier: in some cases it may be found that previous levels of abstraction need to be altered in a 'refactoring' of the model. The experience gained from this analysis can lead to the requirements being changed keeping in mind that abstract modelling serves foremost to detect faults and gaps in the requirements.

The Raven model checker (described in Chapter 10) checks the RIL model against a CTL specification. The CTL specification has to be derived from the modelling process, e.g. from the UML-B model. Verification may fail either due to a violation of the given CTL specification or due to "state explosion". In case of failure, the model and its specification have to be analysed and corrected. Discovery of a violated CTL specification will in many cases be accompanied by a counter-example to support analysis. Failure due to the "state-explosion" problem may be assumed, after a reasonable amount of time has passed.

Once the model has been successfully verified, both functionally and temporally, it can be developed further. Eventually, it can be implemented. Implementation models are written in B0 [1] or BHDL respectively [11]. Using the corresponding translators these can be translated into software or hardware descriptions. In the PUSSEE project the hardware description has been newly developed.

A system may be decomposed into subsystems when it becomes too large and complicated to verify or comprehend. The decomposition tool assists in carrying out decompositions. Allocation of variables and events to subsystems need to be specified, and the corresponding subsystems are generated automatically. The definition of the subsystems can also serve for documentation and further analysis, e.g. number of shared variables, number of shared events, etc.

6.1 U2B translator

The U2B translator converts UML-B models into B components. Translation from UML-B into B is necessary to gain access to other tools in the toolset.

In many respects B components resemble an encapsulation and modularization mechanism suitable for representing a class. A component encapsulates variables that may only be modified by the operations of the component. UML-B models can be constructed to utilize this natural correspondence between UML classes and B components. Component operations may call the operations of other components when a navigable association exists between the corresponding classes. However, to ensure compositionality of proof, B imposes restrictions on the way variables can be modified by other components (even via local operations). Using this first option imposes corresponding restrictions on the relationships between classes. Only hierarchical class association structures can be modelled. A second option translates a complete UML package (i.e. many classes and their relationships) into a single B component. To use this option a stereotype is given to the package depending on which kind of B component is required. The instance, attributes, associations and operations of the classes in the package are represented in B in the same way but are collated into a single component. This option allows unconstrained (non-hierarchical) class relationship structures to be modelled but no operation calling is possible because all operations are within the same B component (a further restriction for proof reasons). However, if we view the operation bodies as declarative specifications of behaviour rather than encompassing design issues such as how that behaviour is allocated to operations throughout the system, this is not a severe restriction on the UML modelling.

Since B language is not object oriented, class instances must be modelled explicitly. Attributes and associations are translated into variables whose type is a function from the class instances to the attribute type or associated class. For example a variable instance class A with attribute x of type X would result in the following B component:

```
MACHINE          A_CLASS
SETS             A_SET
VARIABLES        A, x
INVARIANT        A : POW(A_SET) & x : A --> (X)
INITIALISATIONA := {} || x := {}
...
```

Operation behaviour may be represented textually in a notation based on B's AMN (abstract machine notation) or as a state chart attached to the class or as a simultaneous combination of both.
Further details of the UML-B modelling language, which is a profile of the UML, are given in Chapter 5. A full description of how UML-B models are translated into B by the U2B translator is given in Chapter 6.

6.2 Atelier B

The use of B language for describing and proving system properties between successive model refinement is crucial for the specific approach. The total number of proofs required, even for small systems, is usually big and difficult to handle without the support of an appropriate tool.

For the automation of proving process Atelier B[2] tool is employed. It is mainly composed of static analyzers that include (a) a *type checker*, which is responsible for the syntactic and the semantic verification of a B component, (b) a *B0 checker* which performs verification specific to the B0 language[3] [1] and (c) the *project verification analyzer*, which performs global verifications on the among system components in order to control the overall system architecture.

Additionally, Atelier B includes proof tools that allow formal proof of the successive B model refinements, where the proof obligations required can be proven either automatically or interactively. More specifically, the proof tools available from Atelier B include:

- The automatic generator of proof obligations from the components in B.
- The rule base manager (the rule base includes more than 2200 rules).
- The automatic prover, which discharges automatically most of the proof obligations. It can be set to a power level to compromise between its speed and proof rate.
- The interactive prover, which is used when the automatic prover has failed. The designer is able to guide the prover with commands and additions of new rules.

[2] Atelier B is a trademark of ClearSy S.A.

[3] B0 is a subset of B language, which is used only in implementations to ensure that those can be directly translated in C/C++/Ada.

- The predicate prover, which demonstrates rules added by the user.
- The graphical visualization of the proof tree of a proof obligation.

Finally, Atelier B supports translation of B implementations to C, C^{++} and Ada for the software parts of the system under development.

6.3 Decomposition assistant

The decomposition assistant[4] automates the decomposition process [12,13] outlined in Figure 3-5. Decomposition is precisely the process by which a certain model can be split into various component models in a systematic fashion. As a result, the complexity of the proving process is reduced while the emerging subsystems can be implemented using different technologies. The communication among the components is automatically generated by the decomposition assistant. Communication consistency is based on the exchange of events among the subsystems and can be formally refined until the final communication scheme is reached.

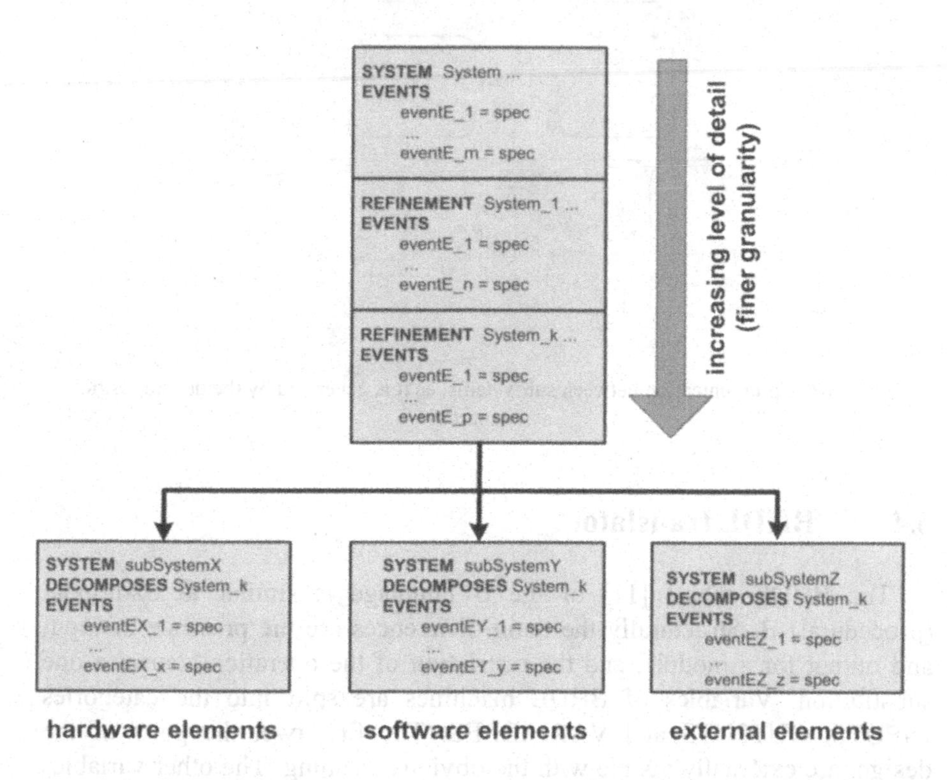

Figure 3-5. System decomposition and hardware/software allocation

[4] The decomposition assistant has been developed by ClearSy S.A. in the context of IST PUSSEE Project [10] .

As described in Figure 3-6, during the decomposition process each variable is allocated to one and only one subsystem and references in other subsystems are copies that need to be updated. The lower part of the figure describes the modules generated by the decomposition assistant: SS1 and SS2. ISS1S2, which contains the communication primitives between the two subsystems, can be further formally refined in order to reach a fully functional protocol implementation. System decomposition [14] is based on a decomposition profile defined by the designer. Decomposition assistant uses the decomposition profile as input and automatically produces the description of each subsystem along with the variables required for describing the selected communication scheme. Reuse of formally proven protocol descriptions through a protocol library is also applicable.

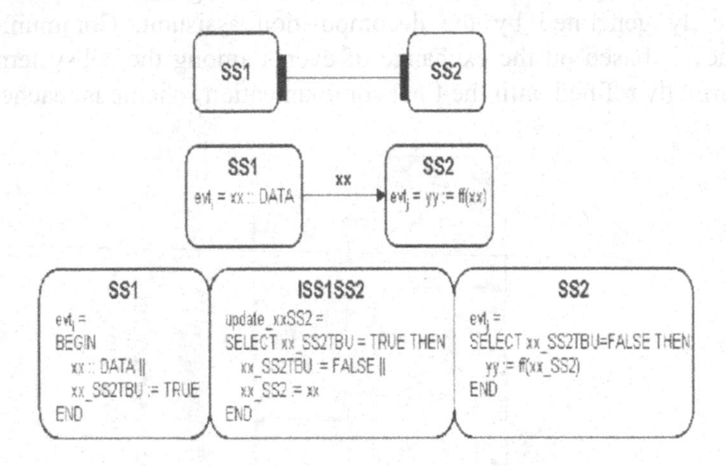

Figure 3-6. Communication between subsystems, as it is generated by the decomposition assistant tool

6.4 BHDL translator

The BHDL subset [11] of the B language is similar to traditional (procedural) B. Structurally the main differences are the presence of input and output for a module, and the restriction of the operation clause to one substitution. Variables of BHDL machines are split into the categories INPUTS, OUTPUTS and VARIABLES. The first two, the ports of the design, are externally visible with the obvious meaning. The other variables are local to the design. The OPERATION clause of a BHDL machine contains a single substitution describing the behaviour of the design. Type and constant declarations may not be made in this kind of BHDL machines.

They must be made in dedicated BHDL machines. This facilitates porting to different target languages.

BHDL data types are restricted to BOOL, INTn and UINTn. In addition, enumeration types can be used and arrays. These types are contained in a basic BHDL machine, called *BHDL.mch* that must be imported in BHDL machines. In SystemC the types correspond to bool, sc_int<n>, sc_uint<n> and enumeration types and arrays. This choice of data types facilitates portability to other hardware description languages.

Arrays are represented as total functions in BHDL and may not be synthesizable after translation. If the design resulting from translation is to be synthesizable, it may be necessary to modify the design first by refinement. For simulation, however, this is not necessary. In practice, we found that most produced VHDL descriptions were readily synthesisable.

The substitution language of BHDL is a subset of the substitution language of B. BHDL machines are cycle accurate models of hardware which can be represented by a design on register transfer level in hardware description languages like VHDL or SystemC. In fact, register transfer level is not strictly enforced because some BHDL constructs and expressions correspond to the behavioural level. This is convenient for simulation.

For performance analysis, a higher abstraction level (and earlier translation in the design process) would be useful. This work is under way for the transaction level of SystemC.

The substitution language of BHDL comprises assignment "x := E", simultaneous substitution "S || T", sequential substitution "S; T", and conditional substitution "IF B THEN S ELSE T END". Arrays can only be assigned as a whole, i.e. assignments of the form "x(k):=E" are not possible. The reason is that tracking intermediary signals created in the translation would be complex and error prone, whereas the price of the incurred restriction in practice is very low. In this article we use lambda expressions to represent array values. So an array assignment has the form: "$x := k.(k \in K|E) \cup k.(k \in L|F)$", where "$dom(x)=K \cup L$". The right-hand side may contain more than two lambda expressions.

Registers are inferred from variables declared in the VARIABLES clause of a BHDL machine depending on their use. Two sets read and write are calculated for a BHDL machine. Inputs declared in the machine must be contained in read, outputs in write. Variables that are contained in read and write are translated into registers. Input and output variables are translated into corresponding ports, and all remaining variables into wires.

Formally the translation into hardware description languages tr(S) of S is based on the before-after predicate $prd_x(S)$ of a substitution S. This is described in Chapter 7.

Identifiers clock and reset must not be used in BHDL machines. Correspondingly named signals for use with registers are produced by the translation. BHDL machine *DELAY* below shows a B implementation of a buffer that delays its input by 8 clock cycles.

```
MACHINE
    DELAY
SEES
    BHDL
INPUTS
    din
OUTPUTS
    dout
VARIABLES
    neg, buffer
INVARIANT
    neg ∈ UINT4 ∧ buffer ∈ 1..8 → UINT4
INITIALISATION
    buffer := λx.( x ∈ 1..8 | 0 )
OPERATION
    BEGIN
        neg := 15 - buffer(1);
        dout := neg
    END ||
    buffer := λx.( x ∈ 1..7 | buffer( x - 1 )) ∪ λx.( x ∈ {8} | din )
END
```

Variable buffer is initialised to an array of zeros. The translation generates a set of registers with reset for variable buffer. The state transition specified by the substitution is translated into a combinatorial circuit. Inputs and outputs are referred to by their original names. The design resulting from the translation is synthesisable. Thus, so is the original BHDL machine *DELAY* from which the VHDL code was produced by automatic translation. The BHDL machine could be further refined, e.g. by implementing the subtraction on bit level. In general, we prefer to allow as many high level constructs as possible, to allow earlier translation and analysis in the development process. For this reason we are also working on a translator to SystemC transaction level.

7. CONCLUSION

We have presented an overview of the PUSSEE method and the associated software tools developed in the EU IST project PUSSEE [10].

The method offers functional and temporal verification in B and RIL, respectively, at high levels of abstraction, but also allows deriving systems from it that are correct by construction. This achieved by means of the B method using formal refinement. Initial system models can be created in UML and verified in B. By formal refinement these are step by step made more concrete towards implementations in the subsets B0 and BHDL of the B language. Models in B0 can be translated into C and BHDL models into SystemC, for instance. Thus the PUSSEE method can be applied to developments that require co-design and (formal) verification.

REFERENCES

1. J-R. Abrial, *The B Book: Assigning programs to meanings*, Cambridge University Press, 1996.
2. J. Warmer, A. Kleppe, *The Object Constraint Language: Precise Modeling with UML*, Addison-Wesley, 1999.
3. P.Facon, R. Lelau, H.P. Nguyen, *Combining UML with the B formal method for the Specification of database applications*, Research Report, CEDRIC Laboratory, Paris, 1999.
4. ClearSy, *Event B Reference Manual. Version 1.0*, Available at: http://www.atelierb.societe.com/ressources/evt2b/eventb_reference_manual.pdf, 2001.
5. J. Draper et al, *Evaluating the B method on an avionics example*, Proceedings of Data Systems in Aerospace (DASIA) Conference, 1996.
6. C. Snook, L. Tsiopoulos, M. Walden, *A Case Study in Requirement Analysis of Control Systems using UML and B*, Proceedings of International Workshop on Refinement of Critical Systems, Methods, Tools and Developments, 2003.
7. J-R. Abrial, *Event Driven Electronic Circuit Construction*, Available at: http://www.atelierb.societe.com/ressources/articles/cir.pdf, 2001.
8. J-L. Boulanger et al, *Formalization of Digital Circuits Using the B Method*, Proceedings of 8th International Conference on Computer Aided Design, Manufacture and Operation in the Railway and Other Advanced Mass Transit Systems, 2002.
9. W. Ifill et al, *The Use of B to Specify, Design and Verify Hardware*, High Integrity Software, Kluwer Academic Publishers, 43-62, 2001.
10. PUSSEE Project. Available at: http://www.keesda.com/pussee, 2004.
11. KeesDA, *BHDL User Guide. Preliminary Version*, Available at http://www.keesda.com.
12. J-R. Abrial, *Event Model Decomposition*, Available at: http://www.atelierb.societe.com/resources/articles/dcmp3.pdf, 2001.
13. T. Lecomte, *D4.4.1: Methodological Guidelines: Interface based synthesis/ refinement in B*, IST-2000-30103 PUSSEE, Project Report, 2003.
14. T. Lecomte, J. R. Abrial, F. Badeau, C. Czernecki, D. Sabatier, C. Snook, *Abstract modeling: System level modelling and refinement in B*, Technical Report, Project IST-2000-30103 PUSSEE, 2003.

Chapter 4

SYSTEM LEVEL MODELLING AND REFINEMENT WITH EVENTB

T. Lecomte
ClearSy, France

1. APPLICATION DOMAIN: DISCRETE SYSTEMS

Event B aims at providing a way to model systems [1][2] that they are made of many parts interacting with a highly evolving (and sometimes hostile) environment. They also quite often involve several concurrent executing agents. They require a high degree of correctness. Finally, most of them are the result of a construction process which is spread over several years and which requires a large and talented team of engineers and technicians.

Although their behavior is certainly ultimately continuous, such systems are most of the time operating in a discrete fashion. This means that their behavior can be faithfully abstracted by a succession of steady states intermixed with « jumps » which make their state suddenly changing to others. Of course, the number of such possible changes is enormous, and they are occurring in a concurrent fashion at an unthinkable frequency. But this number and this high frequency do not change the very nature of the problem: such systems are intrinsically discrete. They fall under the generic name of transition systems. Having said this does not make us moving very much towards a methodology, but it gives us at least a common point of departure.

We already know however that program testing (used as a validation process in almost all programming projects) is by far an incomplete process. Not so much, in fact, because of the impossibility to achieve a total cover of all executing cases. The incompleteness is rather, for us, the consequence of

J. Mermet (ed.), UML-B Specification for Proven Embedded System Design, 53–68.
© 2004 *Kluwer Academic Publishers. Printed in the Netherlands.*

the lack of an oracle which would give, beforehand and independently of the tested object, the expected results of a future testing session. Needless to say that in other more complex cases, the situation is clearly far worse.

It is nevertheless the case that today the basic ingredients for complex system construction still are « a very small design team of smart people, managing an army of implementers, eventually concluding the construction process with a long and heavy testing phase ». And it is a well known fact that the testing cost is at least twice that of the pure development effort. Is this a reasonable attitude in 2002? Our opinion is that a technology using such an approach is still in its infancy. This was the case at the beginning of last century for some technologies, which have now reached a more mature status (for example avionics).

Here the technology we consider is that concerned with the construction of complex discrete systems. As long as the main validation method used is that of testing, we consider that this technology will remain in an underdeveloped state. Testing does not involve any kind of sophisticated reasoning. It rather consists of always postponing any serious thinking during the specification and design phase. The construction of the system will always be re-adapted and re-shaped according to the testing results (trial and error). But, as one knows, it is quite often too late.

In conclusion, testing always gives a short-sighted operational view over the system in construction: that of execution. In other technologies, say again avionics, it is certainly the case that people eventually do test what they are constructing, but the testing is just the routine confirmation of a sophisticated design process rather than a fundamental phase in it. As a matter of fact, most of the reasoning is done before the very construction of the final object. It is performed on various « blue prints » (in the broad sense of the term) by applying on them some well defined practical theories.

2. GLOBAL FRAMEWORK

The B method is based on a hierarchical stepwise refinement and decomposition of a system. After initial informal specification of requirements, an abstraction is made to capture, in a first formal specification, the most essential properties[1] of a system. This top-level abstract specification is made more concrete and more detailed in steps, which may be one of two types. The specification can be refined by changing the data structures used to represent the state information and/or changing the bodies of the operations that act upon these data structures.

[1] For example these could be the main safety properties in a safety critical system

Alternatively, the specification can be decomposed into subsections by writing an implementation step that binds the previous refinement to one or more abstract machines representing the interfaces of the subsections. In a typical B project many levels of refinement and decomposition are used to fully specify the requirements. Once a stage is reached when all the requirements have been expressed formally, further refinement and decomposition steps add implementation decisions until a level of detail is reached where code is automatically generated.

Central to the B method is the concept of full formal verification. At each refinement or decomposition step proof obligations are generated and must be discharged in order to prove that the outputs of the step are a valid refinement of the previous level. At each step when more detailed requirements are introduced or implementation steps are taken, it is proved that they respect all the previous levels. This method ensures that the developed program obeys the properties expressed in all the levels of specification from which it is derived.

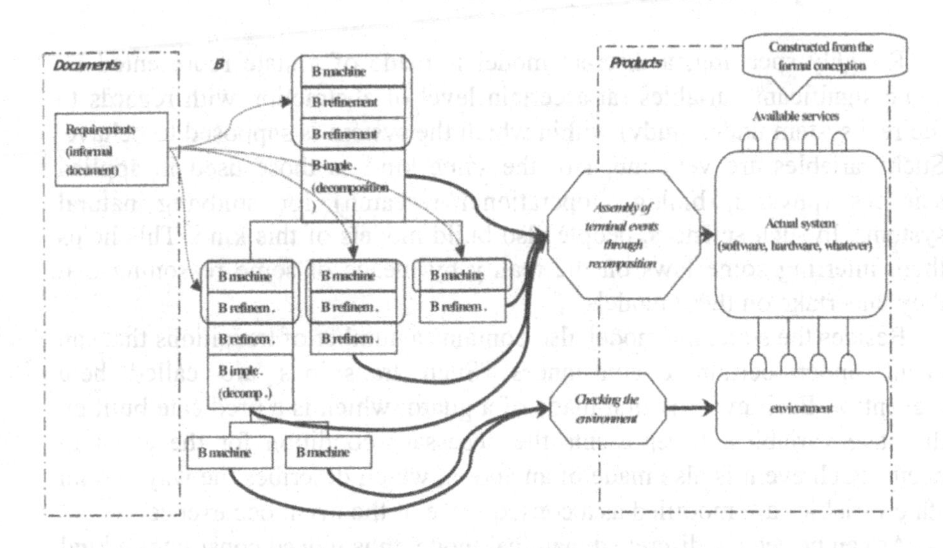

Note that the models we are going to construct will not just describe the « control » part of our intended system. It will also contain a certain representation of the environment. In fact, we shall essentially construct closed models able to exhibit the actions and reactions that take place between a certain environment and a corresponding (possibly distributed) controller, which we intend to construct.

In doing so, we shall be able to plunge the blue print of the controller within (an abstraction of) the environment. The state of such a closed system thus contains « physical » variables (describing the environment state) as well as « logical » variables (describing the controller state). And, in the same way, the transitions will fall into two groups: those concerned by the environment and those concerned by the controller. We shall also probably have to enter into the model the way these two entities communicate.

But as we mentioned earlier, the number of transitions in the real systems under study is certainly enormous. And, needless to say, the number of variables describing the state of such systems is also extremely large. How are we going to practically manage such a complexity? The answer to this question lies in two concepts: refinement and decomposition. It is important to notice here that these two concepts are linked together. As a matter of fact, one refines a model to later decompose it, and, more importantly, one decomposes it to further more freely refine it.

3. MODELING ELEMENTS

3.1 Operational interpretation

Roughly speaking, a discrete model is made of a state represented by some significant variables (at a certain level of abstraction with regards to the real system under study) within which the system is supposed to behave. Such variables are very much of the same kind as those used in applied sciences (physics, biology, operational research) for studying natural systems. In such sciences, people also build models of this kind. This helps them inferring some laws on the reality by means of some reasoning that they undertake on these models.

Besides the state, the model also contains a number of transitions that can occur under certain circumstances. Such transitions are called here « events ». Each event is first made of a guard, which is a predicate built on the state variables. It represents the necessary condition for the event to occur. Each event is also made of an action, which describes the way certain state variables are modified as a consequence of the event occurrence.

As can be seen, a discrete dynamical model thus indeed constitutes a kind state transition « machine ». We can give such a machine an extremely simple operational interpretation. Notice that such an interpretation should not be considered as providing any « semantics » to our models (this will be given later by means of a proof system), it is just given here to support their informal understanding.

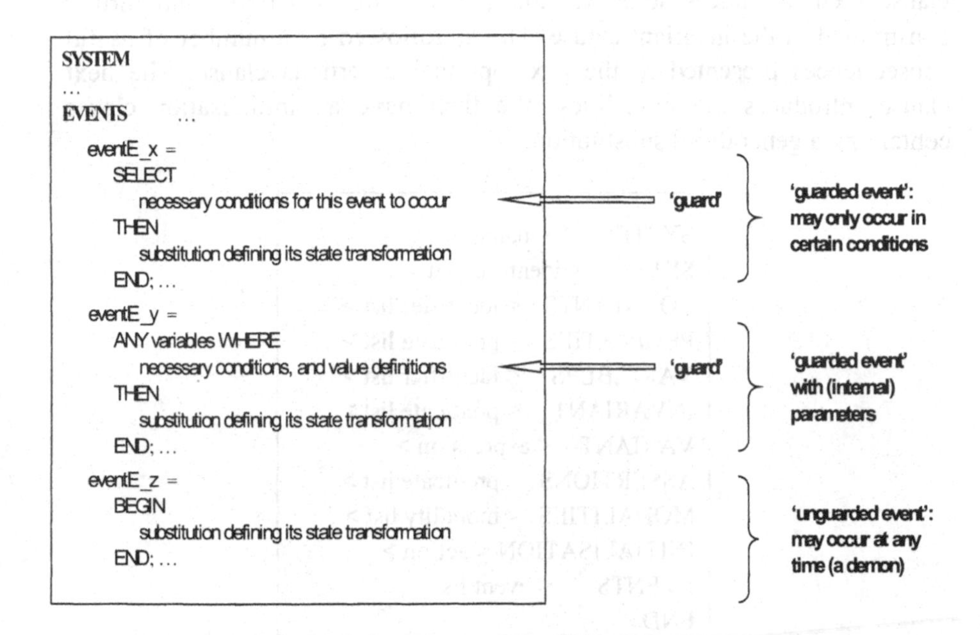

First of all, the « execution » of an event, which describes a certain observable transition of the state variables, is considered to take no time. As an immediate consequence, no two events can occur simultaneously. The « execution » is then the following:

When no event guard is true, then the model execution stops: it is said to have deadlocked.

When some event guards are true, then one of the corresponding events necessarily occurs and the state is modified accordingly, finally the guards are checked again, and so on.

This behavior clearly shows some possible non-determinism (called external non-determinism) as several guards might be true simultaneously. When only one guard at a time is true, the model is said to be deterministic. Note that we make no assumption concerning the specific event which is indeed executed among those whose guards are true. We could at least suppose the simplest assumption, namely that they all have an equal probability to be executed, but this does not add really anything new.

3.2 Discrete model description

A formal model is made of various clauses [5][6]. It first has a name. Then comes an optional sets clause introducing a number of generic sets. This is followed by an optional constants clause introducing some constants, which are typed and further constrained in the properties clause. A variables clause then introduces some variables, which are also typed and further constrained in the invariant clause. This is followed by a number of useful consequences presented in the next optional assertions clause. The next clause introduces the modalities. We then have an initialisation clause containing a generalized substitution.

```
SYSTEM      < name >
SETS        < identifier list >
CONSTANTS   < identifier list >
PROPERTIES  < predicate list >
VARIABLES   < identifier list >
INVARIANT   < predicate list >
VARIANT     < expression >
ASSERTIONS  < predicate list >
MODALITIES  < modality list >
INITIALISATION < action >
EVENTS      < event list >
END
```

3.2.1 Basic Sets

The sets clause contains a list of distinct identifiers. Each of them represents a certain carrier set, which is completely independent from the others. Such abstract sets have no special properties besides being non-empty. They serve as basic types for the formal expressions. They finally represent a first level of possible parameterization.

3.2.2 3.2.2 Constants and their Properties

This constants clause contains a list of constants. Such constants represent a second level of parameterization [4]. The properties clause partially defines these constants in terms of the sets declared in the previous clause. The properties constitute then an axiomatization of the constants. The predicates expressing these axioms are defined within the language of First Order Predicate Calculus with equality extended by Set Theory.

3.2.3 3.2.3 Variables and their Invariants

The variables clause contains a list of state variables.

The invariant clause gives the relevant permanent laws that these variables must follow[2]. Such laws are expressed in terms of the previously defined sets and constants. As for the constants, the predicates expressing these invariants are defined within the language of First Order Predicate Calculus with equality extended by Set Theory.

3.2.4 Event Shapes

The events clause contains a list of events. Each event may be defined in one of the following three forms:

```
name =
BEGIN
    gen:
substitution
END
```

```
name =
SELECT
    condition
THEN
    gen:
substitution
END
```

```
name =
ANY variables
WHERE
    condition
THEN
    gen:
substitution
END
```

where name is an identifier and generalized substitution defines the transition associated with the event. The first event is not guarded: it can be enabled any time. The guard of the other events, which is the necessary condition for these events to occur, is represented by condition in the first case, and by # *variables.condition* in the second one. That second case defines a non-deterministic event where variables represent a list of distinct local variables.

3.2.5 Modalities

We might also consider more complicated forms of reasoning involving conditions which, in contrast with the invariants, do not hold permanently.

[2] An invariant is a condition on the state variables that must hold permanently. In order to prove this, it is just required to demonstrate that under the invariant in question and under the guard of each event, the invariant still holds after being modified according to the transition associated with that event.

The corresponding statements are called modalities[3].
More precisely, we would like to prove that, in certain circumstances (that is, as long as a certain condition C1 holds), another condition C2, which perhaps does not hold now, will certainly hold within a finite future. What we call here a « finite future » only means « after finite occurrences of certain specific events E ».

When such a modality holds, the condition C2 is said to be reachable under the condition C1 and thanks to the events E. More precisely, such a modality formalizes a certain progressing process made of the events E and maintaining the local invariant C1 until the final condition C2 holds. This informal description dictates what is required to prove.

3.2.6 Modalities Concerning a Precise Immediate Future

In this first case, the final condition is to be established by the various events E1 or … or En just after the « execution » of any one of them. The final condition P(x0; x) depends on the values of the current state variables x just after the execution and possibly also on the values x0 of the state variables just before the execution. One can also make precise under which starting condition this eventuality can occur. This yields three different modalities. In the first one, no condition is required. In the second one, the starting condition C(x) only depends on the values of the state variables x just before execution. In the last form, the initial condition C(t; x) depends, besides x, on the values of some variables t that are local to the modality. In that case, the final condition P(t; x0; x) also depend on t.

[3] A modality is a formal statement expressing that a certain « final » condition, which, contrarily to the invariant conditions, does not hold permanently, may nevertheless hold under certain circumstances. There exist two kinds of modalities. They correspond to the fact that the final condition is to be established either in an immediate future or in an unprecise finite future. In both cases, the final condition is to be established by a list of events denoted by E1 or : : : or En. Such a list is made explicit in the modality. In case all current events are concerned by the modality, the keyword all replaces the previous list of events.

```
BEGIN
    E1 or … or En
ESTABLISH
    P(x0; x)
END
```

```
SELECT
    C(x)
THEN
    E1 or … or En
ESTABLISH
    P(x0; x)
END
```

```
ANY t WHERE
    C(t; x)
THEN
    E1 or … or En
ESTABLISH
    P(t; x0; x)
END
```

3.2.7 Modalities Concerning an Imprecise Future

In this second case, the final condition is to be established by the various events E1 or … or En some « time » after the combined « executions » of some of them. Notice that the « time » in question could be null: this means then that the final condition holds initially. Notice that, contrarily to the previous case, the final condition P(x) only depends on the values x of the state variables after the various event executions.

Moreover, as long as the final condition P(x) is not established, an intermediate « current » condition Q(x) must hold. This condition is in fact a local invariant of the process progressing towards the final condition P(x). In order to ensure the termination of this process it is required to exhibit a variant expression V (x), which denotes a natural number.
The process we have just described can also depend on some local variables t constrained by a certain typing condition T(t). When this is the case, then both mentioned conditions as well as the variant may depend on t.

This results in the following two modalities:

```
BEGIN
    E1 or … or En
MAINTAIN
    Q(x)
UNTIL
    P(x)
VARIANT
    V (x)
END
```

```
ANY t WHERE
    T(t)
THEN
    E1 or … or En
MAINTAIN
    Q(t; x)
UNTIL
    P(t; x)
VARIANT
    V (t; x)
END
```

The intermediate conditions Q(x) or Q(t; x) might be missing. In this case the modalities are simplified as follows:

```
BEGIN
        E1 or … or En
LEADSTO
        P(x)
VARIANT
        V (x)
END
```

```
ANY t WHERE
        T(t)
THEN
        E1 or … or En
LEADSTO
        P(t; x)
VARIANT
        V (t; x)
END
```

3.3 Discrete model refinement

Refinement allows us to build a model gradually by making it more and more precise (that is, closer to the reality). In other words, we are not going to build a single model representing once and for all our reality in a flat manner: this is clearly impossible due to the size of the state and the number of its transitions. We are rather going to construct an ordered sequence of embedded models, where each of them is supposed to be a refinement of the one preceding it in that sequence. This means that a refined (more concrete) model will have more variables than its abstraction: such new variables are the consequence of a closer look at our system.

A useful analogy here is that of the scientist looking through a microscope. In doing so, the reality is the same, it does not change, but the look at it is more accurate: some previously invisible parts of the reality are now revealed by the microscope. An even more powerful microscope will reveal more parts, etc. A refined model is thus one that is spatially larger

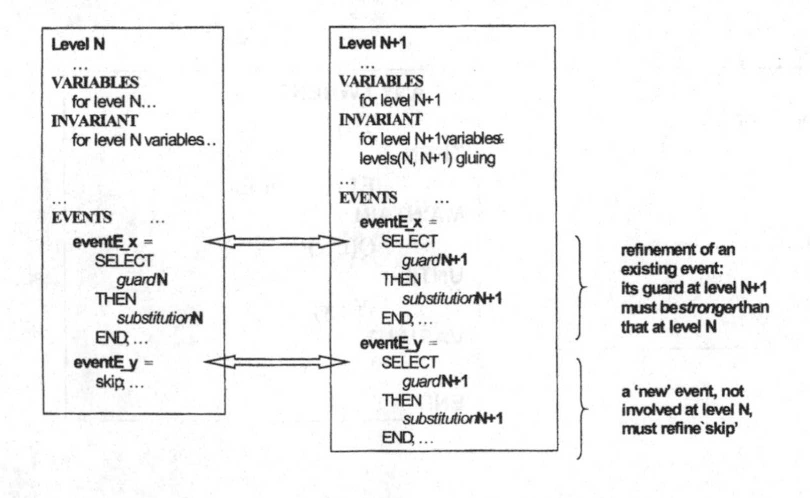

than its previous abstractions. And correlatively to this spatial extension, there is a corresponding temporal extension: this is because the new variables are now able to be modified by some transitions, which could not have been present in the previous abstractions simply because the concerned variables did not exist in them. Practically this is realized by means of new events involving the new variables only (they refine some implicit events doing « nothing » on the abstraction). Refinement will thus result in a discrete observation of our reality, which is now performed using a finer time granularity.

3.3.1 General Form of a Refinement

A refinement is a formal construct very similar to that of a model. As a model, it is made of various clauses. It has a name and it stipulates that it refines a certain model or refinement, which is said to be its corresponding abstraction. It may introduce new sets and constants in the optional sets, constants and properties clauses. These items form new parameterization arguments. The variables clause lists the variables that are kept from the previous abstraction and also those introduced in this refinement. Such new variables are typed and further constrained in the invariant clause. This is followed by an optional variant clause concerning the new events. Next comes some useful consequences presented in the next optional assertions clause. We then have a modality clauses (modalities) and an initialisation clause (initialisation) containing a refined substitution of the more abstract initialisation. A final events clause contains the refinement of the abstract events as well as some new events.

3.3.2 Data and Event Refinements

Data refinement is the process by which some abstract state variables are replaced by some new refined state variables in a refinement. Let x be the set of state variables of an abstraction and y be the set of state variables of its refinement. The two sets are linked by a certain gluing invariant J(x; y), which corresponds to the predicates present in the invariant clause of the refinement.

```
REFINEMENT    < name >
REFINES   < name >
SETS      < identifier list >
CONSTANTS  < identifier list >
PROPERTIES  < predicate list >
VARIABLES  < identifier list >
```

```
INVARIANT  < predicate list >
VARIANT  < expression >
ASSERTIONS < predicate list >
MODALITIES < modality list >
INITIALISATION < action >
EVENTS        < event list >
END
```

3.3.3 Introduction of New Events

New events can be introduced in a refinement. Each of them has to be proved to refine the implicit abstract non-guarded event that does nothing (skip). This means that the new events are only working with the new state variables introduced in the refinement.

The new events, besides refining skip, must also be « confined ». In other words, the collection of new events must not have the possibility to collectively take control for ever over the old events that were already present in the abstraction. Let y be the state variables of the refinement. The confinement can be proved by exhibiting a certain natural number quantity V (y), which must be decreased by each new event. This expression V (y) is introduced in the optional variant clause of a refinement.

3.3.4 Relative Deadlock freeness Preservation

An extra global refinement condition is the one stipulating that the set of all refined events does not deadlock more often than the abstract one. In particular, if the abstract model is live (never deadlocks), we want that a corresponding refinement does not deadlock either. This relative deadlock freeness can be proved by means of the following statement, where the $G_i(x)$ and $H_j(y)$ denote the abstract and refined guards respectively[4]:

$$I(x) \ \& \ J(x, y) \ \& \ G_1(x) \ \& \ \ldots \ \& \ G_n(x) \ y \ H_1(y) \ \& \ \ldots \ \& \ H_m(y)$$

[4] This expression complies with one requirement expressed in "Extensions and enhancements of the existing EventB-method", related to the liveness of event systems: "*In event B it is generally assumed that the disjunction of all guards is true. This may in practice be unnecessarily restrictive. Instead, one could specify liveness predicate P that must imply the disjunction of all guards. In refinements one would require that the disjunction of all guards of the abstraction implies the disjunction of all guards of the refinement. The advantage of using this explicit liveness predicate would be that wanted (or accepted) deadlocks could be distinguished from accidentally introduced ones.*"

I(x) denotes the invariant of the abstraction and J(x,y) the gluing invariant between the current refinement and its abstraction.

3.3.5 Refinements of modalities

Modalities concerning an immediate future are « naturally » implicitly refined to modalities concerning an imprecise future. This is due to the presence of the new events in a refinement: such events will postpone (but not indefinitely as we know) the establishment of the final condition, which was established in an immediate future in the abstraction.

Modalities concerning an imprecise future are clearly refined to modalities of the same kind. Suppose we have an abstract modality with events E1, … , En. The refined modality now corresponds to a progressing process concerned not only by the events F1, …, Fn, which are the refinements of the previous ones, but also by the new events N1, … , Nm present in the refinement. The problem is that the guards of the abstract events E1, … , En are now refined by stronger guards in events F1, … , Fn. This strengthening could have the bad effect of stopping the refined progressing process too early (that is before establishing the final condition). In order to prevent this, one has to prove a certain relative deadlock freeness.

3.4 Discrete model decomposition

Refinement does not solve completely the mastering of the complexity. As a model is more and more refined, the number of its state variables and that of its transitions may augment in such a way that it becomes impossible to manage the whole. At this point, it is necessary to cut our single refined model into several (almost) independent pieces.

Decomposition [3] is precisely the process by which a certain model can be split into various component models in a systematic fashion. In doing so, we reduce the complexity of the whole by studying (refining) each part independently of the others. The very definition of such a decomposition implies that independent refinements of the parts could always be put together again to form a single model that is guaranteed to be a refinement of the original one. This process can be further applied on the components, etc.

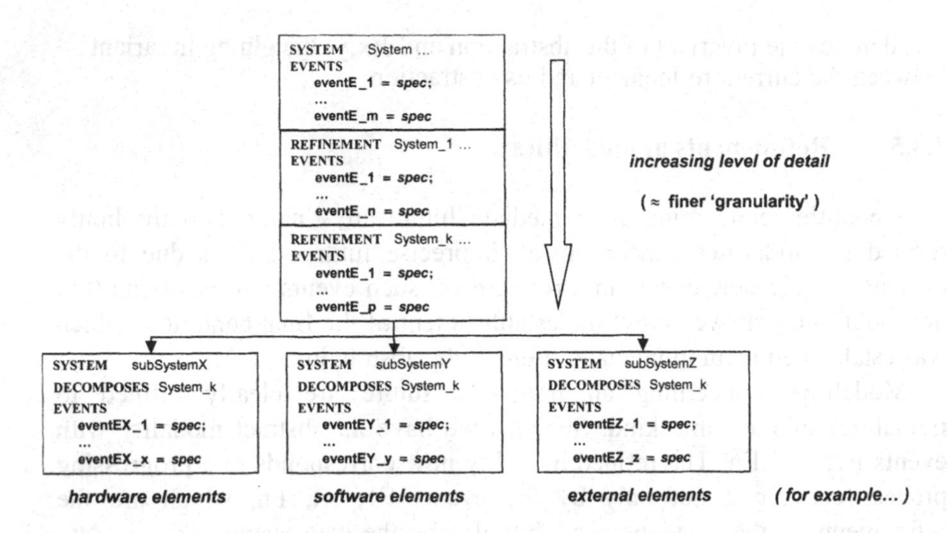

The process of developing an event model by successive refinement steps usually starts with a very few events (sometimes even a single event) dealing with a very few state variables. It usually ends up with many events and many variables. This is because one of the most important mechanisms of this approach consists of introducing new events during refinement steps. The refinement mechanism is also used at the same time to significantly enlarge the number of state variables.

```
SYSTEM          < name >
DECOMPOSES      <name>
SETS            < identifier list >
CONSTANTS       < identifier list >
PROPERTIES      < predicate list >
VARIABLES       < identifier list >
INVARIANT       < predicate list >
VARIANT         < expression >
ASSERTIONS      < predicate list >
MODALITIES      < modality list >
INITIALISATION  < action >
EVENTS          < event list >
END
```

The new events, let us recall, are manifestation of the refinement of the time grain within which we might, more and more accurately, observe and analyse our dynamic system.

At some point, we might have so many events and so many state variables that the refinement process might become quite heavy. And we may also figure out that the refinement steps we are trying to undertake are not involving any more the totality of our system (as was the case at the beginning of the development): only a few variables and events are concerned, the others only playing a passive, but noisy, role.

The idea of model decomposition is thus clearly very attractive: it consists of cutting an heavy event system into smaller pieces which can be handled more comfortably than the whole. More precisely, each piece should be refine-able independently of the others. But, of course, the constraint that must be satisfied by this decomposition is that such independently refined pieces could always (in principle) be easily recomposed. This process should then result in a system which could have been obtained directly without the decomposition, which thus appears to be just a kind of « divide-and-conquer » artefact.

3.4.1 Outcome and Constraints of Decomposition

This decomposition process may play three important practical methodological roles:

1. It is certainly easier (less proofs) to refine (possibly several times) N , . . . , P independently of each other rather than together.
2. This process of decomposition is « monotonic » in that refinements of N , . . . , P can be further decomposed in the same way, and so on.
3. The models N , . . . , P could already possess (off the shelf) some refinements that can then be reused in several projects.

This new decomposition process do not modify in any way the mathematical definition and concept of refinement, but is distinct (although close) from that of importation as described in « classical B ».

When decomposing, the following constraints apply on models:

- all systems decomposing a refined system should have the same events, even with abstracted (simplified) body,
- variables present in the refined system and in more than one decomposing systems can't be refined anymore.

REFERENCES

1. Abrial, J.R: Discrete System Models (2003)
2. Abrial J.R: Guidelines to Formal System Studies (2000)
3. Abrial J.R: Event Model Decomposition (2002)
4. Abrial J.R, Cansell D., Lafitte G.: High Order Mathematics in B (2002)
5. Lecomte T.: Event B Reference Manual (2001)
6. Lecomte T.: Evt2b User Manual (2001)

Chapter 5

THE UML-B PROFILE FOR FORMAL SYSTEMS MODELLING IN UML

Colin Snook[1], Ian Oliver[2] and Michael Butler[1]
[1]University of Southampton, Southampton, UK; [2]Nokia Research Centre, Helsinki, Finland

Abstract: The UML is a popular modelling notation that has a natural appeal to hardware and software engineers and is adaptable through extension mechanisms. Formal (mathematical) modelling languages, on the other hand, are seen as difficult and costly to use and have achieved only limited use despite the benefits that they offer. In previous work, we have proposed an integration of UML and the formal notation, B and provided an automatic translator that produces a B specification. The integrated modelling notation, UML-B, inherits from both UML and B but primarily, is a specialisation of the UML. To achieve this integration we have specialised UML modelling elements via stereotypes, added tagged values to represent B modelling features and imposed constraints to ensure that UML-B models are translated into usable B. Here we describe ongoing work to define UML-B as a profile in accordance with the UML extensibility mechanisms.

1. INTRODUCTION

The UML [4] is successful as a visual modelling notation for the design and communication of object-oriented systems. Formal (mathematical) modelling notations provide verification and validation benefits. By providing a formal interpretation of UML modelling elements and adding behavioural specifications in an integrated format translation to a formal notation is possible. This enables the consistency of a model, internally and between successive refinements, to be formally verified by proof. Also, model checking tools provide an effective, faster and cheaper verification mechanism than proof while still being more rigorous than traditional verification techniques such as reviewing. Animation provides a means of validating specifications prior to implementation. These verification and validation processes are not available in the UML even if annotated

J. Mermet (ed.), UML-B Specification for Proven Embedded System Design, 69–84.
© 2004 *Kluwer Academic Publishers. Printed in the Netherlands.*

constraints are added in the UML constraint language, OCL [5]. Hence the main motivation is to provide a translation from the UML to a recognised formal notation that has good tool support to enable rigorous verification and validation of UML models. However, there is no such formal notation that follows the object-oriented paradigm. Therefore translation from unrestricted UML models is problematic. To overcome this we define a specialisation and enhancement of the UML that is amenable to translation Into the chosen formal notation.

The U2B translation (Chapter 6) translates class diagrams with attached state charts into B [1]. The Class diagram defines the structure of B components and their variables. Further textual information in the specifications of classes and operations defines constraints and operation semantics. This textual information is expressed in a form of the B notation that adopts an object oriented dot style for referencing class instances. State Machines may be attached to classes and used to define the effect of operations on a state variable. Refinements between classes can be represented and a large system can be decomposed using a hierarchy of UML packages. Different styles of model are catered for including a conventional B development or an event or action systems approach.

However, B is not an object oriented modelling language and consequently some UML features are difficult to translate. B contains restrictions in order to achieve its primary goal of decomposition into provable modules. In order to achieve a successful translation, restrictions must be placed on the UML models that can be translated and in many cases UML modelling elements are assigned special meanings. The UML contains extensibility mechanisms for defining specialisations of its notation for use in particular modelling types. The UML profile [3] is the main mechanism for doing this. A UML profile may include the following steps:

1. Identification of the UML subset relevant to the profile.
2. Definition of specialisations using the extension mechanisms of UML (stereotyping and tagged values).
3. Imposition of syntactical rules that restrict the models that can be created. (Well-formedness rules)
4. Definition of a particular semantics.

This paper describes a UML profile, called UML-B, that embodies these restrictions and semantics. The semantics of UML-B are provided by examining the semantics of the equivalent B model produced by the U2B translator.

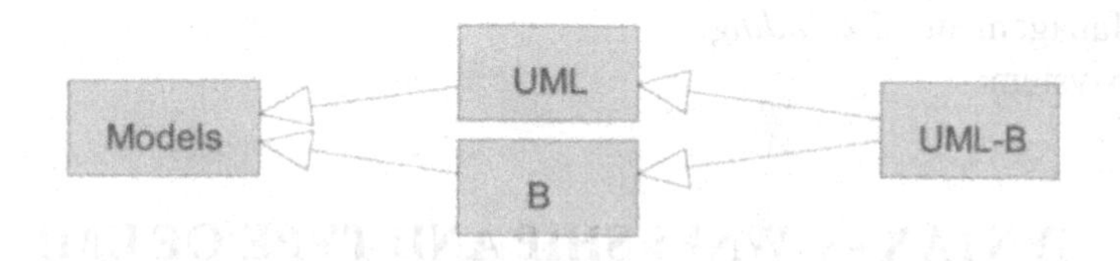

Figure 5-1. classification of model types

UML-B is a class of models. It is a specialisation of the class of models that use the UML notation. It also inherits from the class of models that use the B notation. The U2B translator reads a UML-B model and creates a B model.

2. UML SUBSET RELEVANT TO UML-B.

Since, in UML-B models, full behavioural information is attached to the model elements themselves, the scenario-based notations of UML (collaboration diagrams) are not used. Currently UML-B does not use the use case notation either, though this may be considered in the future. This section identifies the subset of the UML notation upon which UML-B is defined. This is done, first by identifying the relevant packages and excluded classes from the UML metamodel, and secondly by listing the UML model elements. In the latter, we take the opportunity of specifying some constraints on ownership relations between model elements.

The UML-B profile retains the following from the UML Metamodel [3].

Foundation::Core:: *Excluding:*

Method (from Backbone),

Permission and Usage (from Dependencies),

Interface, Node, Component, ElementResidence and Artifact (from Classifiers),

Foundation::ExtensionMechanisms

Foundation::DataTypes *(for use in the metamodel)*

BehaviouralElements::CommonBehaviour *Excluding:*

Signal, Exception and Reception (i.e. all of signals)

CreateAction, ReturnAction, SendAction TerminateAction, UninterpretedAction and DestroyAction (from Actions)

SubSystemInstance, ComponentInstance, NodeInstance (from Instances)

BehaviouralElements::StateMachines*Excluding:*

SynchState, SubState, CompositeState and SubMachineState (from Main)

SignalEvent, TimeEvent,ChangeEvent (from Events)
ModelManagement *Excluding:*
Subsystem

3. SYNTAX - OWNERSHIP AND TYPE OF UML SUBSET MODEL ELEMENTS

UML-B borrows most of its syntax from UML. For example the usual UML restrictions on ownership of other modelling elements and inter-element connectivity apply. However, UML-B requires additional restrictions in terms of cardinality of ownership and the values permitted in text fields. Table 1 lists the retained UML model elements and specifies UML-B constraints on the ownership of one model elements by an other. Where the UML feature is a text field, type information is also given. The indentation indicates ownership of features by other UML features. The number range in the left hand column of the table, headed #, gives the permitted cardinality of the ownership of that feature. For example there must be one and only one package at the top level that owns everything else. This package may have one stereotype. If it does, then the value of that stereotype must be one of the strings listed.

Table 5-1. Ownership cardinality and field type constraints on the use of UML modelling elements in UML-B. (# = cardinality constraints of ownership)

#	UML feature	field type
1..1	package	
1..1	stereotype	procedural system | event system | action system
0..n	package	system | machine | refinement | implementation
0..1	documentation	UML-B clause list
1..1	stereotype	system | machine | refinement | implementation | module
1..n	class	
0..1	documentation	UML-B clause list
1..1	stereotype	system | machine | refinement | implementation |constant
1..1	name	identifier
1..1	cardinality	number range
0..n	parameter	
1..1	name	identifier
1..1	type	type set
0..n	attribute	
1..1	stereotype	static |constant
1..1	name	identifier
1..1	type	type set
0..1	initial value	value | ::set
0..n	operation	

<table>
<tr><th>#</th><th>UML feature</th><th>field type</th></tr>
<tr><td>0..1</td><td>documentation</td><td>comment</td></tr>
<tr><td>1..1</td><td>stereotype</td><td>static |create | destroy</td></tr>
<tr><td>1..1</td><td>name</td><td>identifier</td></tr>
<tr><td>0..1</td><td>return type</td><td>identifier list</td></tr>
<tr><td>0..n</td><td>parameter</td><td></td></tr>
<tr><td>1..1</td><td>name</td><td>identifier</td></tr>
<tr><td>1..1</td><td>type</td><td>type set</td></tr>
<tr><td>0..1</td><td>pre-condition</td><td>UML-B predicate</td></tr>
<tr><td>0..1</td><td>semantics</td><td>UML-B substitution</td></tr>
<tr><td>0..1</td><td>state model</td><td></td></tr>
<tr><td>1..1</td><td>stereotype</td><td>petri</td></tr>
<tr><td>1..1</td><td>name</td><td>identifier</td></tr>
<tr><td>1..n</td><td>state</td><td></td></tr>
<tr><td>0..1</td><td>documentation</td><td>UML-B clause list</td></tr>
<tr><td>1..1</td><td>name</td><td>identifier</td></tr>
<tr><td>0..n</td><td>transition</td><td></td></tr>
<tr><td>1..1</td><td>name</td><td>identifier</td></tr>
<tr><td>0..1</td><td>guard</td><td>UML-B predicate</td></tr>
<tr><td>0..1</td><td>action</td><td>UML-B substitution</td></tr>
<tr><td>1..1</td><td>target</td><td><state> | <decision></td></tr>
<tr><td>0..1</td><td>initial state</td><td></td></tr>
<tr><td>0..n</td><td>transition</td><td>(as in state)</td></tr>
<tr><td>0..1</td><td>final state</td><td></td></tr>
<tr><td>0..n</td><td>transition</td><td>(as in state)</td></tr>
<tr><td>0..n</td><td>decision</td><td></td></tr>
<tr><td>1..n</td><td>transition</td><td>(as in state)</td></tr>
<tr><td>0..n</td><td>association</td><td></td></tr>
<tr><td>1..1</td><td>stereotype</td><td>static | constant | POW | POW1 | seq | seq1</td></tr>
<tr><td>1..1</td><td>client rolename</td><td>identifier</td></tr>
<tr><td>1..1</td><td>client multiplicity</td><td>number range</td></tr>
<tr><td>1..1</td><td>client navigability</td><td>boolean</td></tr>
<tr><td>1..1</td><td>supplier</td><td><class></td></tr>
<tr><td>1..1</td><td>supplier rolename</td><td>identifier</td></tr>
<tr><td>1..1</td><td>supplier multiplicity</td><td>number range</td></tr>
<tr><td>1..1</td><td>supplier navigability</td><td>boolean</td></tr>
<tr><td>0..n</td><td>dependency</td><td></td></tr>
<tr><td>1..1</td><td>stereotype</td><td>sees | includes | imports | refines</td></tr>
<tr><td>1..1</td><td>name</td><td>identifier</td></tr>
<tr><td>1..1</td><td>supplier</td><td><class></td></tr>
<tr><td>0..n</td><td>generalisation</td><td></td></tr>
<tr><td>1..1</td><td>stereotype</td><td>POW | POW1 | seq | seq1</td></tr>
<tr><td>1..1</td><td>supplier</td><td><class></td></tr>
</table>

4. UML-B TAGS

In this section we describe how UML-B specialises standard UML model elements. This is done using the extensibility features (stereotypes and tagged values) of UML. We assume the unification suggested by Fontoura, Pree and Rumpe in their UML-F profile for Framework Architectures [2]. In this unification stereotypes are equivalent to boolean tagged values. The existing stereotype notation, <<stereotypeName>>, may be used but is considered equivalent to a tagged value: {stereotypeName=true}. This enables us to talk about *tags,* which are *tagged values*, some of which may be *boolean* a.k.a. *stereotypes.* In many case, although a tag has been defined in the UML-B profile, the specialisation is derived implicitly from the context of the model element.

4.1 Model Types

The UML-B model can use one of several modelling styles. The intended modelling style is indicated by a tag on the top-level package (*Logical Package*). The translation to B operates differently depending on the modelling style. The available modelling styles are shown in table 2 (note that these tags are only valid when applied to the top level package in a model).

Table 5-2. UML-B tags for model types

Tag Name	Applies to	Type	Description
procedural system	Logical Package	Boolean	The model type is a conventional model where operations are interpreted like procedures.
event system	Logical Package	Boolean	The model represents an event system where operations are interpreted as events and parameters represent non-deterministic internal choice.
action system	Logical Package	Boolean	The model represents an action system where operations are interpreted as actions and parameters represent external choice

4.2 Model Architecture

Packages are used to represent modules or components. Modules are collections of components. Components may be machines, refinements or implementations. (The meaning of machine, refinement and implementation is equivalent to the corresponding B terms). Each module package contains one or more classes or packages. Classes also represent components (machines, refinements or implementations) or elements within an owning component package. Classes also represent an encapsulation of data and operations. Classes may also represent a set of instances in the usual object-oriented sense. The components may have dependency relationships between them that are interpreted in various ways. The tags that are used to distinguish these modelling specialisations are shown in table 3.

Table 5-3. UML-B tags for model structuring and interrelationships

Tag Name	Applies to	Type	Description
module	Package	Boolean	the package represents a B module.
machine	Package	Boolean	the package represents a B module.
refinement	Package	Boolean	the package represents a B module.
implementation	Package	Boolean	the package represents a B module.
machine	Class	Boolean	The class represents a machine component.
refinement	Class	Boolean	The class represents a refinement component.
implementation	Class	Boolean	The class represents an implementation component
includes	Dependency	Boolean	the supplier component is included in the client (equivalent to machine inclusion)
imports	Dependency	Boolean	the supplier component is imported in the client (equivalent to machine importation)
sees	Dependency	Boolean	the supplier component is seen in the client (equivalent to machine sees)
refines	Dependency	Boolean	the client component refines the supplier component (equivalent to refinement in B). (For notation the UML realises arrow is used).
parameters	Dependency	list of identifiers (,)	a list of actual parameters when instantiating a parameterised component (via includes or imports)

4.3 Additional Information in UML-B

The tags in this section provide a mechanism for specifying additional modelling elements that are not catered for by specialisations of UML modelling elements. Most of these tags provide a mechanism for adding textual B clauses to the UML modelling element representing a UML-B component. In these cases, the tag name is the corresponding clause name in B from which the purpose of the tag value can be deduced. (Only the most commonly used clauses are shown here). Characters in brackets in the type column are the separator to be used when the type is a list.

A class cardinality tag is used to specify variations in class instance modelling. If a number range is specified, variable instance modelling is provided including a create operation. However, in many modelling situations a fixed number of objects exist. If the class cardinality is set to a fixed number (i.e. the upper and lower extent of the range are equal, then a constant set of instances is modelled. Since all instances exist from the beginning, there is an implicit initialisation phase where they are assigned initial values. It is also useful to model an implicit single instance. If the cardinality is set to one, no instance modelling is provided simplifying the model.

A tag is provided in order to identify the list of operation return parameters. B infers the type of return parameters from the operation body. Hence the UML operation return type is unused. However, instead, B needs to know which identifiers represent the definition of the returned values.

Table 5-4. UML-B tags for instance cardinality and additional B fields

Tag Name	Applies to	Type	Description
CARDINALITY	Class	number range	cardinality determines the mutability of the class instances set as well as its cardinality.
INSTANCES	Class	an enumerated set or other set expression	definition of instances of the class
CONSTRAINTS	Class	predicate	a predicate that expresses a constraint on the parameters of the component
SETS	Class	identifier list (;)	a list of deferred or enumerated sets
CONSTANTS	Class	identifier list (,)	a list of concrete constants
PROPERTIES	Class	predicate	a predicate that expresses a constraint on the constants of the component
DEFINITIONS	Class	definition list (;)	a list of definitions
VARIABLES	Class	identifier list (,)	a list of variables
INVARIANT	Class	predicate	a predicate that expresses a

Tag Name	Applies to	Type	Description
			constraint on the variables of the component
INITIALISATION	Class	substitution	a substitution that sets the initial value of class variables
INITIALISATION_ PREDICATE	Class	predicate	a predicate that defines the initial value of class variables
results	Operation	identifier list (,)	list of identifiers that indicate values returned by the operation body.

5. UML MODEL ELEMENTS USED WITHOUT TAGS

This section describes how standard UML model elements are used in UML-B.

Package Name - The package name is used as the basis of the component name.

Class Name - The class name is used as the basis of the component name. The class name also represents a set of instances which may be referred to in the behavioural specifications of other classes.

Class Parameters - If a parameterised class is used, the class parameters are interpreted as constants or sets when they are instantiated. The class parameter name is a placeholder for the actual constant or set. The parameter type may be used to specify a set that will be used to constrain the possible values of the parameter.

Class Attributes - Class attributes are variables of the component. The attribute type is a set that constrains the possible values of the variable in an invariant. The attribute may be given an initial value that, for variable instance classes, is the value assigned to the variable in the 'create' operation, and for fixed instance and singular classes, is the value given to the variable at initialisation.

Class Associations - Class associations are variables of the component. The supplier role name is the name of the variable and the type of the variable is a relationship between the instances of the client and supplier classes. The multiplicities of the supplier and client roles in the association represent an additional constraint invariant on the values of the association variable.

Class Operations – Class operations describe possible changes in the state of the class. In an event or action system they represent a description of events or actions that might happen within the system. In this case, apart from describing the change to the state variables of the class, they may also constrain (via guards) the state under which they do and do not occur. Note that an event does not occur unless there is an available choice action where all the guards are true, including those of called operations if applicable. In a procedural system, the operation must not control when it is available and hence there is always an implicit do nothing transition interpreted from state charts if no state transition is unguarded.

Operation Parameters - The operation parameter name is a placeholder for the actual value passed when the operation is invoked. (For event and action systems the parameter represents variations in a family of similar events or actions). The parameter type may be used to specify a set that will be used to constrain the possible values of the parameter.

Operation Pre-conditions – For procedural systems the operation pre-condition describes constraints on the parameters and class variable state in which the operation actions are valid. If the operation is invoked outside of its pre-condition anything can occur. For event or action systems, the pre-condition represents a guard that defines when the operation does and does not occur.

Operation Semantics – The operation semantics describe changes to the state of class variables that occur when the operation is called or when the event/action occurs. (The operation semantics occur in parallel to any state changes defined by a state model).

Class State Model – The class state model also describes changes to the state variables that occur when the operation is called or when the event/action occurs.

State Model Name– For normal (non-petri) state models, the state model name represents an attribute of the class whose value for each instance determines the current state as described in the state model.

State Model States - For normal (non-petri) state models, the state model states are a set of values that the state model attribute can take. For petri state models each state represents a variable that is a subset of the class instances currently in that state.

State Model Transitions – State model transitions represent possible changes to the state model attribute or state instance subsets. For normal (non-petri) state models, the new value of the state model attribute is the target state but the new value can only be assigned when the current value equals the starting state. For petri state models, the instance is removed from the starting state and added to the target state. The state change can only occur when the relevant operation is called or when the event/action occurs.

The relevant operation or event/action is given by the transitions event attribute. The transition may also specify additional guards and actions.

State Model Decision Pseudostates – Decision pseudostates may be used to split a single transition into a choice of alternative transitions and to merge several transitions into a single transition. This is for convenience since the choice/merge can be represented without decision pseudostates as separate transitions. However, the technique has been found useful when refining state models.

6. SUMMARY OF UML-B MAPPING ON TO UML

This section summarises the specialisation of UML model elements in UML-B by listing the UML-B model element and giving the UML model elements and/or tags from which it is derived.

Table 5-5. Summary of mapping from UML-B model elements to standard UML model elements

#	UML-B model element	UML model element and/or UML-B profile tag
1..1	event model	Logical View package <<event system>>
1..1	action model	Logical View package <<action system>>
1..1	procedural model	Logical View package <<procedural system>>
0..n	module	package <<module>>
1..n	machine component	class or package <<machine>>
0..n	refinement component	class or package <<refinement>>
1..1	implementation component	class or package <<implementation>>
1..1	component name	class name or package name
1..1	instance modelling variable or constant	class name
1..1	instance modelling constraint or invariant and initialisation	class {CARDINALITY}
0..1	instance modelling set	class {INSTANCES}
0..n	component parameter	class parameter
1..1	parameter name	parameter name
1..1	parameter type constraint	parameter type
0..n	component variable	class attribute
1..1	variable name	attribute name
1..1	variable type invariant	attribute type
0..1	variable initial value	attribute initial value
0..n	component operation	class operation

#	UML-B model element	UML model element and/or UML-B profile tag
1..1	operation name	operation name
0..1	operation return identifiers	operation {results}
0..n	operation parameter	operation parameter
1..1	parameter name	parameter name
1..1	parameter type pre-condition	parameter type
0..1	operation precondition	operation precondition
0..1	operation body	operation semantics || class state model (see below)
0..n	component action	class operation
1..1	action name	operation name
0..1	action return identifiers	operation {results}
0..n	action parameter	operation parameter
1..1	parameter name	parameter name
1..1	parameter type pre-condition	parameter type
0..1	action guard	operation precondition
0..1	action body	operation semantics || class state model (see below)
0..n	component event	class operation
1..1	event name	operation name
0..n	event local variable	operation parameter
1..1	local variable name	parameter name
1..1	set for variable value choice	parameter type
0..1	event guard	operation precondition
0..1	event body	operation semantics || class state model (see below)
0..1	component variable	class state model <<>>
1..1	state variable name	state model name
1..1	state variable type set and invariant	state model set of states
1..n	type set element	state name
0..1	state variable initial state	state model target of transition from initial state
0..n	(affects interpretation of transitions)	state model decision pseudostate
0..n	state variable changes in operation body (see operation body above)	state model transition
1..1	(locate relevant operation)	transition name
0..1	state variable change	transition guard

#	UML-B model element	UML model element and/or UML-B profile tag
	guard	
0..1	state variable change additional actions	transition action
1..1	state variable new value	transition target
0..1	component variables	class state model <<petri>>
1..n	state variable name	state name
0..1	initialisation: variable=instances	target of transition from initial state
0..n	(affects interpretation of transitions)	state model decision pseudostate
0..n	state variable changes in operation body (see operation body above)	state model transition
1..1	(locate relevant operation)	transition name
0..1	state variable change guard	transition guard
0..1	state variable change additional actions	transition action
1..1	new location for instance	transition target
0..n	component variable mapping instances	class association
1..1	mapping variable name	association supplier role name
1..1	mapping invariant (type)	association supplier
1..1	mapping invariant (range)	association client role multiplicity
1..1	mapping invariant (domain)	association supplier role multiplicity
0..n	component includes other component	class dependency <<includes>>
1..1	includes other component	dependency supplier
0..1	includes rename	dependency name
	includes actual parameters	dependency {parameters}
0..n	component imports	class dependency <<imports>>
1..1	imports other component	dependency supplier
0..1	imports rename	dependency name
	imports actual parameters	dependency {parameters}
0..n	component sees	class dependency <<sees>>
1..1	sees other component	dependency supplier
0..1	component refines	class dependency <<refines>>

#	UML-B model element	UML model element and/or UML-B profile tag
1..1	refined component	dependency supplier
0..1	component inherits	class generalisation
1..1	inheriting component	generalisation client
1..1	inherited component	generalisation supplier
0..1	component textual B clauses	class documentation

7. WELL-FORMEDNESS RULES

This section lists rules that must be followed when constructing UML-B models.

1. If the 'prefix class name' option is not used, identifiers must be unique throughout a model. An identifier is a string of two or more characters from a..z, A..Z, 0..9, _ The first character must be from a..z, A..Z. Due to instance modelling naming conventions, class name identifiers must not contain characters from a..z. An identifier list is a comma-separated list of identifiers.
2. Number Ranges must be an expression of the form min..max giving the set of numbers which are permitted. Min or max can be left indeterminate by setting to either n or *. For example, 0..n = any number, 1..n = any number but at least one, 1..1 = exactly one, 5..5 = exactly five
3. Type Sets must be an expression resulting in a set that is visible to the class. For example, any valid B base type (e.g. NAT), the name of another class, an expression involving other attributes/associations of the class (e.g. union, function), a set (POW, FIN etc) of any of these, a sequence (seq, iseq etc) of any of these.
4. Attribute initial values are optional but if given must either be a value from the type set or "::" followed immediately by a B expression giving a subset of the type set. If the latter form is used the variable will be non-deterministically assigned one of the values from the subset.
5. All Relationships must be binary
6. Only association relationships can loop back to the same class
7. Only sees and imports relationships can be between classes in different packages
8. Associations must be navigable in (only) one direction.
9. Each module package must own exactly one root machine (a root machine is a class that is not the supplier of any intra-package association or dependency and not the client of any refines relationship).
10. In module packages, the structure of intra-package relationships (excluding sees relationships) between classes must be a tree.

11. In module packages, visibility is restricted to elements owned by a component class and those of classes that it has a navigable association path to.
12. In module packages, multiple associations between the same pair of component classes must be in the same direction
13. UML-B predicates and substitutions are equivalent to their B counterparts except that they use the dot notation for instance referencing.
14. In module packages, UML-B substitutions must not contain simultaneous calls to operations of another component (whether via the same or different associations).
15. In module packages, UML-B substitutions must not contain calls to operations within that component (even if the class has an association to itself).
16. In component packages, no operation calling is permitted (but components have full visibility within the package).
17. A class cannot be the client of a refines relationship and the supplier of an association.
18. Implementations cannot be the client or supplier of an association
19. The chain of refines relationships within a package must be a linear sequence from an implementation or refinement (the most concrete specification) to a root machine (the most abstract specification).
20. An implementation class must be the client of a refines relationship and must not be the supplier of a refines relationship
21. If a class has parameters the only relationships it can be the supplier for, are includes relationships.
22. Includes relationships to a parameterised class instantiate the supplier class and must have the following format for the relationship name: rename(actual1,actual2….) , where rename is an optional new name for the instantiated class (to distinguish it and its namespace from other instantiations) and the list of actual parameters matches the formal parameters of the instantiated class (and are of valid type).
23. Between each pair of states that has a connecting sequence of transitions (via decision points if sequence is more than one transition), there must exist exactly one transition that is named and that name matches an operation of the component.
24. Operation stereotypes <<create and <<destroy>> can only be used by varying instance classes.
25. Multiple inheritance is not permitted.
26. INITIALISATION and INITIALISATION_PREDICATE tags cannot both be used in the same model.

8. CONCLUSIONS

The UML-B profile defines a specialisation of UML models with additional tags that enable the translation of UML-B models into the B notation. The B form allows rigorous verification and validation of UML-B models. While it is necessary to place restrictions on the kind of UML models that can be translated, the essential benefits of the UML are retained, especially if the changes envisaged in current work are successful.

REFERENCES

1. Abrial, J.R.: *The B Book - Assigning programs to meanings*. Cambridge University Press. (1996)
2. Fontoura, M., Pree, W. & Rumpe, B.: UML-F profile for Framework architectures. Addison Wesley (2002)
3. OMG Manual: http://www.omg.org
4. Rumbaugh, J., Jacobson, I. & Booch, G.: The Unified Modelling Language Reference Manual. Addison-Wesley. (1998)
5. Warmer, J. and Kleppe, A.: The Object Constraint Language: precise modelling with UML. Addison-Wesley. (1999)

Chapter 6

U2B

A tool for translating UML-B models into B

Colin Snook and Michael Butler
University of Southampton, Southampton, United Kingdom

Abstract: The UML is a popular modelling notation that has a natural appeal to hardware and software engineers and is adaptable through extension mechanisms. Formal (mathematical) modelling languages, on the other hand, are seen as difficult and costly to use and have achieved only limited use despite the benefits that they offer. In previous work, we have proposed an integration of UML and the formal notation, B called UML-B. The integrated modelling notation, UML-B, inherits from both UML and B but primarily, is a specialisation of the UML. Here we describe how UML-B can be used to obtain a useful translation into B using the U2B translation tool.

1. INTRODUCTION

In the previous chapter we introduced a UML profile, UML-B, that defines a formal modelling notation based on UML [4] and some features borrowed from B. One of the main motivations for this profile was to enable formal verification and validation using a translation into the B notation [1]. This chapter describes the use of UML-B and the translation into B using the U2B translator [7,8,9,10]. Firstly we explain some of the features of B that make translation from an object-oriented modelling language such as UML, difficult. Then we describe the translation performed by the U2B tool. This is done by explaining how UML features can be used in accordance with the UML-B profile and how these features are translated into B. The U2B tool utilises the extensibility features of Rational Rose [3].

J. Mermet (ed.), UML-B Specification for Proven Embedded System Design, 85–108.
© 2004 *Kluwer Academic Publishers. Printed in the Netherlands.*

2. FEATURES OF THE B LANGUAGE THAT MAKE TRANSLATION FROM UML MODELS DIFFICULT

Some features of the B language make it difficult to map object-oriented models, such as those created in the UML-B, to B. These features are, in general, due to the main goal of B, which is to facilitate modular proof of large systems. The difficulties they present are not insurmountable but mean that the development of a useful UML_B to B translation is challenging. The main motivation for translating UML-B into B (rather than a free-er specification language such as Z) is to enable design refinements to be formally proven. Therefore, we note that necessary conditions for a translation to be useful is that the B is reasonably natural and does not complicate the proof process. This attention to ease of proof is a major distinguishing feature compared to similar and previous work. However, some of the basic translation details are similar to and borrow from other work such as [6].

2.1 Non-object-oriented-ness

Firstly, the most obvious difficulty is that B is not object oriented. The most basic feature of object orientation is the ability to model classes of objects via abstract data types. B has an encapsulation mechanism (machines) that allows variables to be grouped with the operations that act upon them. It is also possible, via machine renaming, to instantiate several instances of a machine. However, there is no mechanism to use a B machine to specify the behaviour of an indeterminate or variable set of instances. For example, Z has 'promotion' which enables schemas to be used to define a behaviour that is then promoted and bound to a set of instances at a higher level. This limitation can be overcome by explicitly modelling the set of instances within the machine and modelling each feature of the type by a function whose domain is the set that models the instances. This is what the U2B translator does.

2.2 Restrictions on B component and variable access

B contains restrictions on the way that operations can be called between and within machines, and on simultaneous changes to machine variables. These restrictions are necessary in order to achieve or simplify compositionality of proof. The restrictions have different implications, listed in the subsections below, depending on which UML entities are represented within a B component. The restrictions are as follows:

a) A machine cannot have more than one other machine that makes calls to its operations. This restriction prevents data sharing involving multiple write access.
b) There must not be any loops within the calling structure of a set of machines.
c) Operations cannot call other operations within the same machine.
d) Simultaneous calls to several operations of another machine are not allowed.
e) Each variable of a machine can be altered by, at most, one of the simultaneous substitutions of an operation
f) A variable of a machine can only be written by the operations of the machine that contains it

If a translation is used where each class is represented by a B component, the B restrictions above lead to similar restrictions imposed on the relationships between classes. That is, for class relationships that allow operation calls (such as associations), only strictly hierarchical relationship structures are permitted. Associations cannot be altered by both the associated end classes since this would require write sharing of a variable.

The restriction on calling operations within the class can be avoided by repeating the actions of the 'called' operation within the 'calling' operation in place of the call. (The disadvantages of repeating blocks of substitutions can be avoided by using B definitions. Note, however, that the repetition generates separate proof obligations for each copy).

Operation semantics where more than one instance of an associated class is modified simultaneously cannot be translated to valid B if the one to one mapping between class methods and machine operations is maintained. This can be overcome by constructing a single operation which alters the attribute values for multiple instances in a single substitution.

Operation semantics where more than one instance of the class is modified simultaneously cannot be done by parallel alterations of the attribute value of each instance (as would be natural in an object-oriented style). This can be overcome by altering the function mappings for both instances in a single B substitution. E.g. the UML-B expression,

```
ii.att:=ii.att+1 || jj.att:=jj.att-1
```

must be translated to the single B substitution

```
att:=att <+ {(ii |-> att(ii)+1), (jj |-> att(jj)-1)}
```

If a translation is used where all classes are represented in a single B component, the restrictions on class relationships are overcome because full write access to all the variables in the model are available and hence write sharing is available. However, the encapsulation provided by classes is lost and no operation calling is permitted. (The lack of operation calling can be overcome to some extent using definitions to represent called class methods).

3. TRANSLATION RULES OF U2B

The U2B translator converts UML Class diagrams, including attached statecharts, into the B notation. Some entities on the class diagram and statecharts can have UML-B stereotypes to specialise their translation into B. Some entities may have UML-B 'clauses' in which additional modelling entities or constraints can be defined textually. In some cases the pre-defined UML text fields attached to entities are used to textually specify parts of the model in a notation similar to B. This chapter describes the translation mapping from UML to B.

The translator can operate in several modes selected by a stereotype on the top level ('Logical View') package. The modes are procedural system, action system and event system [11]. The methods described in this book utilise the event systems approach. A conventional procedural model represents a subsystem that models an interface with its environment as a set of operations. These operations must always be available since the model has no control over the environment and hence when it will invoke the operations. In event systems, the environment and the subsystems are contained within the same model. Operations are replaced by events that occur within a closed system. Events are not called but may include constraints on when they occur and hence may block. For event systems, the translator uses the Event B notation which has the following differences from conventional, procedural B. The heading, `MACHINE`, is replaced by `SYSTEM`, the `OPERATIONS` clause is replaced by the `EVENTS` clause, additional clauses (`VARIANTS` and `MODALITIES`) are supported. In event systems modelling, pre-conditions are not used (since there is no concept of an external client that must meet a condition). Instead guards may be specified to determine when the event can and cannot occur (a `SELECT` substitution is used). Event parameters are interpreted differently from operation parameters (since there is no concept of an external client that will provide a value). Instead operation parameters indicate a family of related events that differ only in the non-deterministically selected parameter values within the event (an `ANY` substitution is used to select any value from the parameter type. Note that the `ANY` substitution blocks as a guard if there are no values in the type) Events may also be guarded if there is no available transition in a statechart. (Whereas, for procedural models, an '`ELSE skip`' branch is added to all SELECT substitutions that are generated from statecharts).

3.1 Mapping to B Components

The translation allows a choice of mappings to B components. In many respects B components resemble an encapsulation and modularisation mechanism suitable for representing a class. A component encapsulates variables that may only be modified by the operations of the component. The first option utilises this correspondence between UML classes and B components. Each class is translated into a separate component. Component operations may call the operations of other components when a navigable association exists between the corresponding classes. However, as explained above, B imposes restrictions on the way variables of components can be modified. Using this first option, imposes corresponding restrictions on the relationships between classes. Only hierarchical class association structures can be modelled. A second option translates a complete UML package (i.e. many classes and their relationships) into a single B component. The instance, attributes, associations and operations of the classes in the package are represented in B in the same way but are collated into a single component. This option allows unconstrained (non-hierarchical) class relationship structures to be modelled but introduces a new restriction. No operation calling is possible because all operations are within the same B component. Since the operation bodies are intended to be declarative specifications, and the variables of all classes are accessible, this is not a severe restriction on the UML modelling. In the UML model the operation specifications are written in the context of the class to which they belong. Hence variables of another class should only be accessed by navigating a sequence of navigable associations or where a dependency between the classes exists. (Currently this is not enforced by the U2B translation and an illegal inter-class variable access would be undetected).

The translator defaults to the first (class-component) option. To use the second option the stereotype of a package should be set to <<system>>, <<machine>>, <<refinement>> or <<implementation>> depending on which kind of B component is required. When using this package-component option, sub-packages, if any, are included into the component via the INCLUDES clause.

3.1.1 Classes

A class to B component translation is supported by default if the package owning the class is not a component package. (A component package is a package that has one of the following stereotypes: machine, refinement, implementation). The class represents an encapsulation and visibility of other classes data and operations is only possible when a relationship exists between the two classes. Note that it is possible to mix both translation methods. For example, it might be useful to use a single class to represent the most abstract level within a module package (and hence define the interface of the package in terms of externally callable operations) and refine this with a refinement package containing a class diagram.

3.1.2 Packages

The UML 'package' represents a collation mechanism for grouping class diagram modelling entities (such as classes and other sub-packages) into a namespace. Packages, therefore, control visibility of other entities, but do not imply additional semantics such as representation of a type exhibited by instances. In many ways packages are similar to the concept of B components, possibly more so than classes. We therefore use packages in our UML models to represent three levels of model structure. Firstly, the top level package represents a complete system containing all its levels of abstraction and decomposition. Packages contained within the top-level package can represent two further sub-levels; a module containing all its levels of abstraction or a component which represents a single level of abstraction within a system or module. (A class is an alternative means to represent a component).

To distinguish the intended meaning of a package stereotypes are attached to packages. The following are possible stereotypes to be applied to packages in a UML-B model:

Table 6-1. Package stereotypes

Level	Stereotype	translation
system	procedural system	complete B conventional system
system	event system	complete B event system
system	action system	complete B action system
module	module	B module
component	system	B system (event systems only)
component	machine	B machine
component	refinement	B refinement
component	implementation	B implementation

Relationships between packages can either be explicitly defined via a clause in the package documentation or implied from the package ownership hierarchy. If a machine package, Ma, is owned by another component package, it will be translated as the component includes Ma via an 'INCLUDES' clause. If a module package, Mo, is owned by another component package, it will be translated into the component includes the root level machine of Mo via an 'IMPORTS' clause.

3.2 UML-B clauses

UML-B clauses provide a way to add extra modelling information to the UML model, that cannot be expressed diagrammatically. The clauses are written in the specification documentation window of the modelling entity to which they apply. (The clause is a tagged value of the entity but for convenience it is currently embedded in the text value of the tagged value, 'documentation'). The UML-B profile defines the clauses that can be used via tagged values in this way. In fact, any valid B clause (except OPERATIONS which must be represented as described elsewhere in this chapter) has a corresponding meaning in UML-B (some clauses can not be used with some modelling entities). For example, we use this method to specify invariants for a class. Each clause must be headed by its B clause name in capitals and starting at the beginning of a line. For many clauses the text that follows the clause name, up until the next clause title (if any) will just be added to the appropriate clause in the B component. In addition to the usual B clauses, UML-B includes some clauses that extend UML to make modelling options in B available. The additional clauses (CARDINALITY, INSTANCES & INITIALISATION_PREDICATE) are described in more detail later in this chapter.

Where clauses can contain more than one item a clause separator is defined (for B clauses the B separator is used). Any text before the first clause is treated as comment and added as such at the top of the machine.

If a package-component mapping is used, the package documentation field may be used for adding clauses. The information in these clauses should apply to the complete package or to several packages and not to a particular class. For example, a package with stereotype <<refinement>> should have a REFINES clause to specify which package is refined.

3.3 Class instances

Class instances are modelled as a set. Different options for modelling class instances are described in this section but all (except abstract classes which have no instances) result in a variable or constant that represents the set of instances. This set is named after the class name.

The instances set is used in the modelling of the data entities owned by the class. Attributes and associations are translated into variables whose type is defined as a function from the current instances to the attribute type or associated class.

For example consider the following class diagram with classes A and B, where A has an attribute x and there is a unidirectional association from A to B with role y and 0..1 multiplicity at the target end. A second association, w, has a 1..n multiplicity:

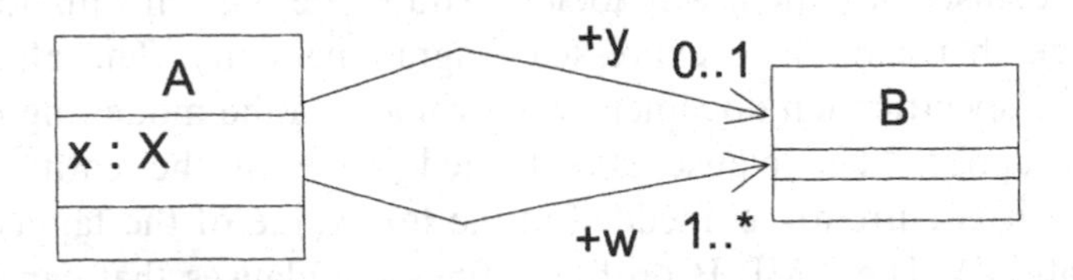

Figure 6-1. Classes with associations

If a class to B component translation is used, this will result in the following machine representing all instances of A:

```
MACHINE          A_CLASS
EXTENDS          B_CLASS
SETS             A
VARIABLES        w, y, x
INVARIANT        w : A --> POW1(B) & y : A +-> B & x : A --> (X)
INITIALISATION   w := {} || y := {} || x := {}
END
```

Note that the multiplicity of the association w is handled as a function from instances of class A to sets of instances of class B using the POW (powerset) operator. Multiplicities of associations are discussed in more detail later. The machine is initialised with no instances and hence all attribute and association functions are empty. A separate machine will be generated for class B.

In the example above, as well as in the examples that follow, we use the usual B conventions for capitalisation of names. That is, type sets, including given or enumerated sets, are named in upper case and variables are named in lower case. Hence attributes and association roles are named in lower case. Class names are given in upper case since they are used to generate the name for the given set of possible instances of the class. This results in the variable representing the set of possible instances being upper case, however this reflects its main role as a type specifier.

3.3.1 Class static entities and static referencing

A stereotype <<static>> may be attached to an attribute, association or operation. This means that the attribute, association or operation belongs to the class rather than a specific instance of the class. This will change the translation so that attributes and associations are typed with no instance mapping and operations have no instance parameter.

Sometimes it is useful to make a static reference to a non-static variable. This can be done by preceding the variable name with $. This inhibits U2B from translating the variable reference into a function application. For example a boolean attribute, `ba`, could be initialised for a class with 3 instances by the clause:

```
INITIALISATION $ba:={i1|->TRUE, i2|->FALSE, i3|->FALSE}
```

(To some extent this requires knowledge of the outcome of the U2B translation but is provided for convenience until better ways of achieving the required models are discovered).

3.3.2 Current Instance

As is normal in UML, the current instance (that is the subject of an operation call) is implicit in UML-B and automatically modelled in the B. However, it can be referenced within the operation's pre-conditions, guards, actions and semantics by using the reserved word 'self'. However, to ensure that the reference is unique throughout the B model, the U2B translator refers to the current instance as '`thisCLASS`' where CLASS is the class name. U2B converts all occurrences of '`self`' to this form. (Alternatively the form '`thisCLASS`' may be used directly instead of '`self`' if preferred).

3.3.3 Class cardinality

Class cardinality is defined using a UML-B clause, `CARDINALITY i..j`, where i and j are integers or 'n'. This is a constraint on the number of instances of a class and is used to determine the type of instance modelling in the corresponding B machine as described below. The current number of instances at any time must be greater than or equal to i and less than or equal to j. If i = j, the number of instances cannot be changed and this affects the modelling of instances as described below. 'n' may be used to indicate that the constraint is unspecified. For example, 0..n is a varying set of instances with no upper limit; n..n is a fixed set of unspecified size. (If i=j the range can be abbreviated to i).

The class cardinality type is determined from the cardinality clause as follows:

0..0 - abstract class with no instance modelling
1..1 - singular class with instance set={`self`}
x..x - fixed instance class with instance set =1..x or a user supplied enumeration (where x is some integer >0)
n..n - fixed instance class with instance set deferred or a user supplied enumeration
x1..x2 - variable instance class with instance set is a variable subset of a deferred set or user supplied enumeration. (where x1/=x2)

3.3.4 Instances clause and instances associations

The instances clause may be used to define a set to be used to model the class instances. (e.g. a set enumeration can be given). It is often useful to define one class' instances based on another

`INSTANCES <set>` where `<set>` is any valid B set definition.

For example:-

```
INSTANCES {sweden,france,england}
INSTANCES 3..6
INSTANCES seq(OTHERCLASS)
INSTANCES CLASSA*CLASSB
```

A more visual alternative for associating one classes instances with those of another is to show it as an association with role name `INSTANCES`. Stereotypes such as <<seq>> and <<iseq>> can be used with an instances association to define the client class' instances as a compound of the client instances.

3.3.5 Varying instance classes

For varying instance classes, class instances are modelled as a set of all possible instances of the class and a variable that represents the current instances of the class as a subset of the possible instances. The set of possible instances can be defined explicitly via a clause in the documentation field of the class (see section on clauses). If no instance set is defined the possible instances set will be modelled as a deferred set.

```
MACHINE           A_CLASS
SETS              A_SET
VARIABLES         A, x
INVARIANT         A : POW(A_SET) & x : A --> (X)
INITIALISATION    A := {} || x := {}
...........
```

3.3.6 Instance creation and destruction

Instance creation and destruction is only enabled for variable instance classes (i.e. when class cardinality is not a range). When this is the case, the standard UML operation stereotypes, <<create>> and <<destroy>> can be attached to operations. If the class has a statechart and its initial transition(s) are named then their actions (currently guards on initial transitions are not supported) will be added to a create operation with the same name (an operation will be added to the class automatically if it does not exist). Similarly, if final transitions are named they will have their guards and actions added to a destroy operation of the same name. A create operation picks any instance that isn't already in use, adds it to the current instances set, and adds a maplet to each of the attribute and association relations mapping the new instance to the appropriate initial value, if given. Alternatively, the operation body can be used to textually specify the initialisation of class variables, possibly using a initial value supplied as a parameter of the operation. A destroy operation removes the instance (identified by the automatically generated operation parameter) from the current instances. The example below is similar to the first example but class `A` has an additional attribute, `z`, that has an initial value, `k`.

```
Return <-- new_A =
PRE    A /= A_SET
THEN
   ANY new WHERE new : A_SET - A  THEN
      A := A \/ {new} ||
      ANY xx WHERE xx:POW1(B) THEN w(new) :=xx END ||
      ANY xx WHERE xx:B THEN y(new) :=xx END ||
      ANY xx WHERE xx:X THEN x(new):=xx END ||
      z(new):=k ||
      Return := new
   END
END
```

Attribute `x` has no initial value specified and is therefore initialised non-deterministically to any value of the type `X`. Attribute `z` is initialised to the specified initial value, `k`. Association `w` must be initialised to a non empty set because its multiplicity may be greater than one but is definitely greater than zero. It is initialised non-deterministically to any non-empty subset of instances of `B`. The association, `y`, is initialised non-deterministically to any instance of `B`.

3.3.7 Static classes

Often it is useful to be able to disable modelling of class instances. For example the specification models a single implicit generic instance of an entity, rather than an explicit set of instances. The resulting specification is simpler and clearer for not modelling instances. If the class cardinality is set to 0..0 in the UML class specification, the U2B translator creates a machine with no instance modelling. Note that this can only be done at the top level of an association hierarchy since at lower levels the instance set is used for referencing by the higher level. Below is shown the machine representing class A from the first example above if the class' cardinality is set to 0..0. Note that there is no modelling of instances; the types of attributes are simpler because it is no longer necessary to map from instances to the attribute type. There can be no instance create operation, attributes are initialised in the machine initialisation clause.

```
MACHINE          A
EXTENDS          B_CLASS
VARIABLES        w, y, x
INVARIANT        w : POW1(B) & y : B & x : X
INITIALISATION   w :: POW1(B) || y :: B || x :: X
END
```

Another situation where class instances should be inhibited is where the class is merely a container for some static definitions or facilities. An alternative method for disabling instances is to make the class a class utility. (This is a standard UML class kind for this situation). Class utilities are treated by U2B in exactly the same way as classes with cardinality 0, but may be a more appropriate modelling technique for this situation.

3.3.8 Fixed instance classes

A third type of instance cardinality is possible. Many examples of modelling embedded real time systems require classes with a small, fixed number of instances. An option for specifying a fixed class cardinality (instead of a range) was introduced. In this case current instances are modelled using a constant set (instead of a variable subset) and instance creation/destruction is disabled. There are three options for modelling fixed instances, as follows.

Deferred instances - If the class cardinality is given literally as n..n (or n for short) the instances are modelled as a deferred set.

Enumerated instances – as for variable instance classes, an INSTANCES clause can be used to specify an enumerated set of instances.

Numbered instances – If neither of the above are true (i.e. the cardinality is i..i, where I is an integer>1 and no INSTANCES clause is given) then the instances are modelled as a constant, C, with the property C=1..i, where C is the class name.

Table 6-2. Instance modeling options. (C is the class name, i is the class cardinality).

	Variable	Deferred	Enumerated	Numbered
SETS	C_SET	C	C={e1,e2,..ei}	
CONSTANTS				C
PROPERTIES				C=1..i
VARIABLES	C			
INVARIANT	C:POW(C_SET)			

3.3.9 Singular classes

Singular classes are a special case of fixed instance classes when the cardinality is 1..1. In this case, the instances set is defined as {thisCLASSNAME}, overriding any INSTANCES definition. U2B translates as for fixed instances classes except that no instance parameter is added to operations and no universal quantification over instances is added to invariants and assertions

3.3.10 Inheritance

Inheritance indicates a sub typing of a class. The instances of the subclass are also instances of the superclass. Hence the modelling of instances is changed in the following ways when a class inherits from a superclass. Note that some combinations of instance modelling types in the subclass and superclass may not be valid. For example, if the subclass is specified as an enumerated set, the set must be a subset of the superclass's instances and hence the superclass could not be modelled as a numbered set of instances. In fact, U2B will automatically add subclass instances to the superclass' instances set if they are not all ready included.

Variable instances – The possible instances set is not needed. Instead the current instances variable is defined as a subset of the current instances set of the superclass.

Deferred instances – The deferred instances set is not needed. Instead an instances constant is defined as a subset of the current instances set of the superclass.

Enumerated instances – The enumerated instances set is not needed. Instead an instances constant is defined as being the given instance set, which must be a subset of the current instances of the superclass.

Numbered instances – As normal, a numbered instances constant is defined accordingly to class cardinality, but the cardinality must be less than or equal to that of the superclass.

Table 6-3. Instance modelling for inheriting sub-classes.(S is the super class name)

	Variable	Deferred	Enumerated	Numbered
SETS				
CONSTANTS		C	C	C
PROPERTIES		C:POW(P)	C={e1,e2,..ej}	C=1..j
VARIABLES	C			
INVARIANT	C:POW(P)			

3.4 Class Data

This section describes the translation of data entities belonging to a class.

3.4.1 Attributes

Attribute types may be any valid B expression that defines a set. This includes predefined types (such as `NAT`, `NAT1`, `BOOL` and `STRING`) functions, sequences, powersets, instances of another class (referenced by the class name), and enumerated or deferred sets defined in the class specification documentation window.

3.4.2 Associations

Uni-directional associations will be translated into functions. The function will be named after the rolename at the target (arrowed) end. In UML, multiplicity ranges constrain associations. The multiplicities are equivalent to the usual mathematical categorisations of functions: partial, total, injective, surjective and their combinations. Note that the multiplicity at the target end of the association (class `B` in the example above) specifies the number of instances of `B` that instances of the source end (class `A`) can map to and vice versa. This can be confusing when thinking in terms of functions because the constraint is at the opposite end of the association to the set it is constraining. The multiplicity of an association determines its modelling as shown in Table 4.1. U2B translates unidirectional associations into functions to sets of the target class instances (e.g. `POW(B)`) to avoid non-functions.

Table 6-4. How associations are represented in B for each possible multiplicity constraint

Ai and Bi are the current instances sets of class A and B respectively and f is a function representing the association (i.e. the role name of the association with respect to the source class, A).
`disjoint(f)` *is defined in B as:*

```
!(a1,a2).( a1:dom(f) & a2:dom(f) & a1/=a2 => f(a1)/\f(a2)={} )
```

UML association multiplicity	Informal description of B representation	B invariant
0..n → 0..1	partial function to Bi	Ai +-> Bi
0..n → 1..1	total function to Bi	Ai --> Bi
0..n → 0..n	total function to subsets of Bi	Ai --> POW(Bi)
0..n → 1..n	total function to non-empty subsets of Bi	Ai --> POW1(Bi)
0..1 → 0..1	partial injection to Bi	Ai >+> Bi
0..1 → 1..1	total injection to Bi	Ai >-> Bi
0..1 → 0..n	total function to subsets of Bi which don't intersect	Ai --> POW(Bi) & disjoint(f)
0..1 → 1..n	total function to non-empty subsets of Bi which don't intersect	Ai --> POW1(Bi) & disjoint(f)
1..n → 0..1	partial surjection to Bi	Ai +->> Bi
1..n → 1..1	total surjection to Bi	Ai -->> Bi
1..n → 0..n	total function to subsets of Bi which cover Bi	Ai --> POW(Bi) & union(ran(f))= Bi
1..n → 1..n	total function to non-empty subsets of Bi which cover Bi	Ai --> POW1(Bi) & union(ran(f))= Bi
1..1 → 0..1	partial bijection to Bi	Ai >+>> Bi
1..1 → 1..1	total bijection to Bi	Ai >->> Bi
1..1 → 0..n	total function to subsets of Bi which cover Bi without intersecting	Ai --> POW(Bi) & union(ran(f))= Bi & disjoint(f)
1..1 → 1..n	total function to non-empty subsets of Bi which cover Bi without intersecting	Ai --> POW1(Bi) & union(ran(f))= Bi & disjoint(f)

3.4.3 State variables

If a class has an attached statechart, a variable or variables will be generated to model the current state of each instance of the class. This is described in the section on statechart behavioural specification.

3.4.4 Constants

A stereotype, <<constant>> may be attached to an attribute or association. This will change the translation so that the attribute or association is translated into a B constant with associated properties instead of a variable and invariant.

The <<constant>> stereotype may also be attached to a class. This has the same effect as attaching the stereotype to all the attributes and associations owned by the class.

3.5 Behaviour

The dynamic behaviour modelled on a class diagram that is converted to B by U2B is embodied in the behaviour specification of class operations and invariants. UML does not impose any particular notation for these definitions; they could be described in natural language or using UML's Object Constraint Language [5]. However since we wish to end up with a B specification it makes sense to use B notation to specify these constraints. The constraints are specified in a notation that is close to B notation but needs to observe a few conventions in order for it to become valid B within the context of the machine produced by U2B. When writing these bits of B the writer shouldn't need to consider how the translation would represent the features (associations, attributes and operations) of the classes. Also when modelling in UML it is more natural to follow the object-oriented conventions of implicit self-referencing and the use of the dot notation for explicit instance references. This is illustrated in examples below.

3.5.1 Invariant

We include invariants as a UML-B clause attached to the entity it constrains. Invariant clauses are most commonly attached to classes, in which case they are generally of two kinds, instance invariants (describing properties that hold between the attributes and relationships within a single instance) and class invariants (describing properties that hold between different instances). For instance invariants, in keeping with the implicit self-reference style of UML, we chose to allow the explicit reference to 'this instance' to be omitted. U2B will add the universal quantification over all instances of the class automatically. For class invariants, the quantification over instances is an integral part of the property and must be given explicitly. Hence, U2B will not need to add quantification and instance references.

For example, if `bx: NAT` is an attribute of class `B` having initial value 0, then the following invariant could be defined in the documentation box for class `B`:

```
bx < 100 & !(b1,b2).((b1:B & b2:B & b1/=b2)=> (b1.bx/=b2.bx)
```

This would be translated to:

```
!(thisB).(thisB:B => (bx(thisB) < 100)) &
!(b1,b2).((b1:B & b2:B & b1/=b2)=> (bx(b1)/=bx(b2)))
```

The translation has added a universal quantification, `!(thisB)`, over all instances of `B` in the first part of the invariant. It is not used in the second part where the invariant already references instances of class `B`.

Invariant clauses may also be attached to packages to enforce constraints between several classes. They may also be attached to states within a state diagram when they constrain other variables of the class while the class is in a particular state. In this case, the implication (`statevar=state =>`) is added automatically by U2B.

The invariant is defined in a definition so that it can be referred to if needed in the specification. The following invariant definitions are created for a package-component.

- type_invariant – invariant defining the instances of all classes and the types of all variables in the component
- classA_invariant – invariant defining any explicit invariant for a particular class, A
- package_invariant – invariant defining any explicit invariant for the whole package.
- invariant==type_invariant & classA_invariant & … & classN_invariant & package_invariant

3.5.2 Initialisation

Two alternative styles of initialisation are supported by U2B; substitution and predicate style.

3.5.2.1 Substitution style initialisation

In this style of initialisation a B `INITIALISATION` clause is constructed composed from the parallel assignments of any initial values of attributes and/or statechart variables specified in the model and any explicit substitutions found in `INITIALISATION` clauses. If a variable is not initialised in the UML-B model by one of these methods then it will not be initialised in the corresponding B component, leading to an error when the B is verified by verification tools.

For example, if the following class, `AA`, has attributes `xx:NAT=0`, `bb:BOOL` and `yy:NAT` and has an initialisation clause, `bb:=TRUE`, the resulting initialisation would be:

```
INITIALISATION xx := AA * {0} || bb:=AA*{TRUE}
```

Note that yy has not been initialised because no initial value has been given for it.

3.5.2.2 Predicate style initialisation

In this style of initialisation, a B `INITIALISATION` clause is constructed in the following form:

<varlist> :(invariant & <initial value constraints>)

where <varlist> is a comma separated list of all the variables owned by the component, invariant is a definition containing the complete invariant of the component and <initial value constraints> is the conjunct of equates to initial values of attributes and/or statechart variables and any other explicit constraints found in `INITIALISATION_PREDICATE` clauses of the component.

For example, if the class `AA` above has an initialisation predicate clause stating that initially `bb=TRUE`, the resulting initialisation would be:

```
INITIALISATION xx, bb, yy :(invariant & xx=AA*{0} & bb=AA*{TRUE} )
```

Note that yy is initialised to any value that satisfies the invariant (i.e. any value of its type if there are no additional invariant constraints).

3.5.3 Operation semantics

Operation guards and actions may be specified in a textual format attached to the operation within the class, or in a statechart attached to the class. Operation behaviour may be specified completely by textual annotation, completely by statechart transitions, or by a combination of both composed as simultaneous specification. Specification by statechart is covered in the next section. This section describes operation textual behavioural specifications.

3.5.3.1 Operation textual behaviour specification

Operations need to know which instance of the class they are to work on. This is implicit in the class diagram. The translation adds a parameter `thisCLASS` of type CLASSinstances (where CLASS is the class name) to each operation. This is used as the instance parameter in each reference to an attribute or association of the class.

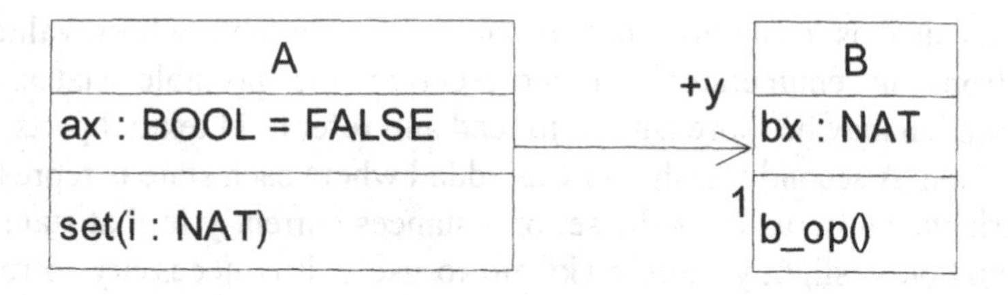

Figure 6-2. Example behaviour

In the above example, the operation `set` might have the guard:

```
i > y.bx
```

and semantics

```
y.b_op() ||
IF y.bx <100 THEN  ax := FALSE ELSE ax := TRUE END
```

which would be translated to

```
i > bx(y(thisA))
```

and

```
b_op(y(thisA)) ||
IF bx(y(thisA)) <100 THEN ax(thisA) := FALSE
ELSE ax(thisA) := TRUE END
```

3.5.3.2 Statechart behavioural specification

For classes that have a strong concept of state change, a statechart representation of behaviour is appropriate. In UML a statechart can be attached to a class to describe its behaviour. The underlying model representing the statechart is constructed and viewed via a set of one or more state diagrams. A statechart consists of a set of states and a set of transitions that represent the changes between states that are allowed. If a statechart is attached to a class the U2B translator combines the behaviour it describes with any operation semantics described in the operation specification semantics windows. Hence operation behaviour can be defined either in the operation semantics window or in a statechart for the class or in a combination of both.

Two possible translations of statecharts are available. The translations are equivalent but use different B data representations. The first translation represents state as a function from instances to a variable whose values are taken from an enumerated set representing the possible states. This representation may be more natural to read and refer to in textual parts of the specification. A second translation was added where each state is represented by a variable whose value is the set of instances currently in that state. This representation is slightly more awkward to use if it is necessary to refer to state in textual parts of the specification but is easier for the proof tools to handle (a higher degree of automatic proof is achieved). The first translation is the default. The second translation is selected if the statechart has the stereotype <<petri>>.

The name of the statechart model is used to define a state variable. (Note that this is not the name of a state diagram, several diagrams could be used to draw the statechart of a class). The collection of states in the statechart is used to define an enumerated set that is used in the type invariant of the state variable. The state variable is equivalent to an attribute of the class and may be referenced elsewhere in the class and by other classes. Statechart transitions define which operation call causes the state variable to change from the source state to the target state, i.e., an operation is only allowed when the state variable equals a state from which there is a transition associated with that operation. To associate a transition with an operation, the transition's name must be given the same name as the operation. Additional guard conditions can be attached to a transition to further constrain when it can take place. All transitions cause the implicit action of changing the state variable from the source state to the target state. (The source and target state may be the same). Additional actions (defined in B) can also be attached to transitions. The translator finds all transitions associated with an operation and compiles a SELECT substitution of the following form:

```
SELECT   statevar(self)=sourcestate1 & transition1_guards
THEN  statevar(self):=targetstate1 || transition1_actions
WHEN  statevar(self)=sourcestate2 & transition2_guards
THEN  statevar(self):=targetstate2 || transition2_actions
   etc
END
```

For a <<petri>> statechart the SELECT substitution would look like this:

```
SELECT self:=sourcestate1 & transition1_guards
THEN     targetstate1:=targetstate1\/{self} ||
         sourcestate1:=sourcestate1-{self} || transition1_actions
WHEN  self:sourcestate2 & transition2_guards
THEN     targetstate2:=targetstate2\/{self} ||
         sourcestate2:=sourcestate2-{self} || transition2_actions
   etc
END
```

The state variable (for each instance) is initialised to the state that is the target of a transition from the initial state. For <<petri>> statecharts the variable representing the initial target state is initialised to the instances set and all others are initialised to empty. For the remainder of this description the non-petri translation is used in examples, although either translation is equally valid.

The `SELECT` substitution generated from the statechart is composed with the operation precondition and body (if any) from the textual specification in the operation's precondition and semantics windows. Actions should be specified on state transitions when the action is specific to that state transition. Where the action is the same for all that operation's state transitions, it may be specified in the operation semantics window in order to avoid repetition.

The following example illustrates how a statechart can be used to guard operations and define their actions. It also shows how common actions can be defined in the operation semantics window.

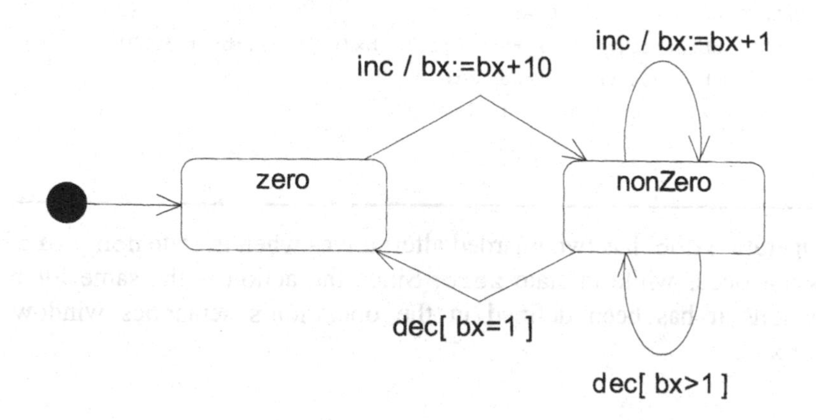

Figure 6-3. Example of statechart behavioural description

The statechart has 2 states, `zero` and `non_zero`. The implicit state variable, `bb_state` (the name of the statechart) is treated like an attribute of type `BB_STATE = {zero,non_zero}`. An invariant, `(bb_state=zero) <=> (bx=0)`, defines the correspondence between the value of the attribute `bx` and the state, `zero`. For all instances, `bb_state` is initialised to `zero` because there is a transition from an 'initial' state to `zero`. The attribute bx has the initial value 0.

```
MACHINE        BB_CLASS
SETS           BB;    BB_STATE={zero,nonZero}
DEFINITIONS
invariant == (
   bb_state : BB --> BB_STATE & bx : BB --> NAT &
   !(thisBB).(thisBB:BB=> ((bb_state(thisBB)=zero)<=>(bx(thisBB)=0)))
)
VARIABLES              bb_state,   bx
INVARIANT              invariant
INITIALISATION         bb_state, bx :(invariant &
                       bb_state = BB * {zero}  & bx = BB * {0} ))
```

Operation `inc` can occur in either state. Its action is different depending on the starting state and so actions have been defined on the transitions and are combined with the state change action.

```
inc =
ANY thisBB WHERE  thisBB:BB THEN
   SELECT bb_state(thisBB)=zero
   THEN   bb_state(thisBB):=nonZero || bx(thisBB):=bx(thisBB)+10
   WHEN   bb_state(thisBB)=nonZero
   THEN   bx(thisBB):=bx(thisBB)+1
   END
END
```

Operation `dec` has two guarded alternatives when in state `non_zero` but does not occur while in state `zero`. Since the action is the same for both transitions it has been defined in the operation's semantics window as `bx:=bx-1`.

```
dec =
ANY thisBB WHERE  thisBB:BB THEN
    SELECT bb_state(thisBB)=nonZero &   bx(thisBB)>1
    THEN   skip
    WHEN   bb_state(thisBB)=nonZero &   bx(thisBB)=1
    THEN   bb_state(thisBB):=zero
    END ||
    bx(thisBB):=bx(thisBB)-1
END
```

3.6 Refinement

The B method is based on a hierarchical stepwise refinement and decomposition of a problem. After initial informal specification of requirements, an abstraction is made to capture, in a first formal specification, the most essential properties of a system. For example these could be the main safety properties in a safety critical system. This top-level abstract specification is made more concrete and more detailed in refinement steps. The specification is refined by changing the data structures used to represent the state information and/or changing the bodies of the operations that act upon these data structures. Central to the B method is the concept of full formal verification. At each refinement step proof obligations are generated and must be discharged in order to prove that the outputs of the step are a valid refinement of the previous level. At each step when more detailed requirements are introduced it is proved that they respect all the previous levels. This method ensures that the developed program obeys the properties expressed in all the levels of specification from which it is derived. Such proof is not easily achieved. While the tool automatically discharges most proof obligations, typically some 20% require human interaction [2]. This interactive proof takes the majority of the effort in a B project development. The form and style of the formal B specification can greatly affect the ease of achieving these proof obligations. Hence ease of proof rather than any design paradigm becomes the primary criterion for developing specifications in B.

We use the realises relationship of UML to indicate that the realising class represents a refinement of the realised class. (When using the package-machine translation mode, a `REFINES` clause is attached to the package). The B component produced from a class, `C_Rn+1`, that realises a class, `C_Rn`, will be a refinement instead of a machine and will have a '`REFINES C_Rn`' clause. If variables in `C_Rn+1` have the same names as those in `C_Rn`, B will assume that they are the corresponding refined versions of the abstract variables. Note however, that the UML representation of the variable could be very different in the realising class from that in the realised class. For example, the state variable defined by a statechart could be the refinement of a class attribute. Where the variables have different names (and possibly not a one to one refinement relationship) a gluing invariant must be specified in the realising class. This must be given in the INVARIANT clause of the class. Note that the realising class may have associations to the same classes as the realised class.

In B refinements, the parameter list must match that of the corresponding operation in the refined component but the preconditions of the refined component are assumed to apply and do not need re-stating. U2B does not therefore generate type preconditions for parameter types in refinements.

REFERENCES

1. Abrial, J.R. (1996) *The B Book - Assigning Programs to Meanings*. Cambridge University Press, ISBN 0-521-49619-5
2.. ClearSy, Aix-en-Provence (F). *AtelierB, Training Course Level 2*
3.. Rational (2000) Rose *Extensibility User's Guide – Rational Rose 2000e*. Rational Software Corporation. Part Number 800-023328-000
4. Rumbaugh, J., Jacobson, I. & Booch, G. (1998) *The Unified Modelling Language Reference Manual*. Addison-Wesley, . ISBN 0-201-30998-X
5. Warmer, J. and Kleppe, A. (1999) *The Object Constraint Language: Precise Modeling with UML*. Addison-Wesley ISBN 0-201-37940-6
6. Meyer, E. & Souquieres, J. (1999) A Systematic approach to Transform OMT Diagrams to a B specification. *FM'99* LNCS1708 1, 875-895
7. MATISSE (2001) Methodology of Integration of Formal Methods Within the Healthcare Case Study. *Deliverable D7 within the Matisse project*, IST-1999-11435, October 2001.
8. Snook, C. and Butler, M. (2000) Verifying Dynamic Properties of UML Models by Translation to the B Language and Toolkit. *In Proceedings of Dynamic Behaviour in UML Models: Semantic Questions. UML 2000 Workshop*
9. Snook, C. (2002) *Exploring the Barriers to Formal Specification* PhD Thesis University of Southampton
10. Snook, C. and Butler, M. U2B - *UML to B translation tool and Manual* V3.5 available at http://www.ecs.soton.ac.uk/~cfs/U2Bdownloads/U2Bdownloads.htm
11. MATISSE (2001) *Event B Reference Manual*. Matisse Project (IST-1999-11435) June 2001 available at http://www.atelierb.societe.com/evt2b/evt2b.html

Chapter 7

BHDL:
Principles and tools for generating proven hardware

S. Hallerstede
KeesDA.SA

Abstract: In this chapter we discuss modeling of hardware and translation to VHDL. Translation to SystemC or Verilog is similar. However VHDL is easier to read and we use VHDL synthesis tools. Translation is important to provide a complete path from formal models to a circuit. Equally important we need a refinement method to arrive at a formal circuit description that can be translated. This method has some significant differences to the refinement method for software. As one would expect, they are virtually not present at system level but become more and more visible as an actual implementation is approached. This means that the initial refinement steps used in hardware are, in principle, also applicable to software, and vice versa.

The subset of the B-language that serves to describe hardware is called BHDL. The definition of the BHDL subset is oriented at the register transfer level for hardware description.
Key words:

1. SYSTEM-LEVEL MODELING

On the system level we use EventB and UML-B to specify abstract system properties. These are refined stepwise until we eventually reach a level where they represent wires and registers (see Figure 7-1).

J. Mermet (ed.), UML-B Specification for Proven Embedded System Design, 109–120.
© 2004 *Kluwer Academic Publishers. Printed in the Netherlands.*

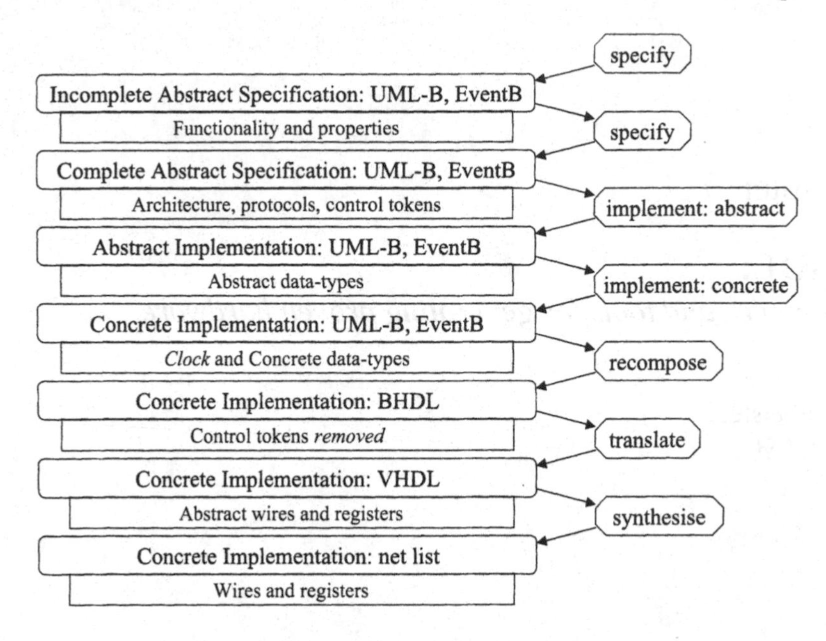

Figure 7-1. EventB/UML-B hardware-design flow

At the most abstract level only some functionality (usually incomplete and to be filled in during refinement) and basic properties (to be completed during refinement) are specified. During the first refinement steps this is completed until all features of the specification have been captured. This concerns mostly functional architecture and protocols. The events are synchronized by passing control token (similar to Petri nets – see the Petri-stereotype in UML-B). From this level we continue to implement the system abstractly. This means mostly that the clock is not introduced as long as possible, and data-type are kept abstract. At some point in the refinement we need to suppose that there is a clock in the system. This clock is not modeled explicitly but represented by the fact that every component in the system must do something if one of them does something. When the clock is introduced, usually also more concrete data-types must be used. During the next refinement steps local state of components is modeled, all data-types are modeled so that they are directly (via translation) implementable, and inputs and outputs are determined. Event recomposition is then used to produce a BHDL description of the hardware. The code that was distributed over a collection of events is put into a single event. By doing this the control tokens are removed, and thus the synchronization that was explicitly specified before is implemented. A BHDL hardware description can be translated into VHDL using the BHDL translator. The resulting VHDL can be synthesized using standard synthesis programs commercially available.

The remainder of this chapter deals with BHDL, the translation to VHDL, and the meaning of BHDL as modeling language for hardware.

2. SUMMARY

The BHDL-language can be described, similar to other subsets of the B-language, in three parts: its machine syntax, basic data-types, and the substitution language. The machine syntax specifies a single operation with input and output variables. The substitution language is particularly suited for hardware description, and the basic data-types facilitate easy translation to various target languages. We lay out the principle of the translation from BHDL to VHD. An optimized version of this translation is implemented in the translator that we use in practice, and the examples presented. The optimizations carried out by the translator reduce the size of the design produced in terms of number of signals, number of blocks, etc. At present, implemented target languages for the translator are VHDL and SystemC. The BHDL language allows translation to all target languages (without modification of BHDL models to suit the target language).

3. MACHINE SYNTAX

The BHDL-machine syntax can be used to describe the behavior of a circuit, or data-types and constants used somewhere in a design. For these purposes two different syntactical forms of machines exist. We refer to the first as operation machine and the second as type machine. It is not possible to declare constants in an operation machine. Neither is it possible to declare variables in a type machine. The reason for this is that translation to different target languages may cause problems when these kinds of declarations are not kept apart.

4. OPERATION MACHINES

An operation machine specifies the behavior of a circuit. It may see a number of type machines specified in the SEES-clause. Input variables and output variables are specified in the corresponding clauses INPUTS and OUTPUTS. In the VARIABLES-clause variables local to the circuit name are specified. From these, registers may be inferred by the translation. The INVARIANT-clause has the usual meaning. The INITIALISATION clause specifies the reset-behavior of the circuit. The OPERATION-clause specifies the behavior of the circuit in terms of one substitution.

```
MACHINE | REFINEMENT | IMPLEMENTATION
      name
REFINES
      name
SEES
      machine_name_list
INPUTS
      variable_list
OUTPUTS
      variable_list
VARIABLES
      variable_list
INVARIANT
      variable_types
INITIALISATION
      substitution
OPERATION
      substitution
END
```

We say a variable is written if it appears on the left hand side of an assignment, and it is read if it occurs on the right hand side of an assignment. Special constraints hold for inputs and outputs of a machine. It may never be written to inputs, and it is necessary to write to an output before it can be read. The initialization of a variable must have one of the following two forms:

x := E ,

where E is an expression, and E does not refer to any variables. If a variable does not need an initial value,

x :∈ TT

may be written in the initialization. The type TT must be the type of variable x. Input variables and output variables must be initialized non-deterministically in the latter form.

5. TYPE MACHINES

A type machine contains declarations of sets and integer constants. After translation, in most programming languages sets become types. Hence, the name. A type machine may see a number of other type machines specified in the SEES-clause. Type and integer constant declarations are made in the SETS and CONSTANTS clauses.

6. DATA-TYPES

The basic data-types available in BHDL have been chosen to match with basic SystemC types, like bool, sc_int<*n*>, sc_uint<*n*>. Their BHDL counterparts are BOOL, INT*n*, and UINT*n* respectively. These types are contained in a basic BHDL-machine BHDL.mch. The choice of these data-types has been made because they are easily portable to other hardware description languages. In addition, enumeration types and arrays can be used. Arrays are represented as total functions in BHDL and are usually not synthesizable after translation. If the design resulting from translation were to be synthesizable, all arrays would have to be removed by refinement first. For simulation, however, this is not necessary.

Sub-ranges of integers may be declared for use as index sets of arrays or in multiple array assignments:

RR = j . . k .

If a BHDL-machine is to be translated to SystemC and a sub-range RR = j . . k is to be used as array index in an array declaration, then j must be 0.

Using the predefined types above arrays may be defined:

AA = RR → TT ,

where RR is a sub-range and TT is predefined type from BHDL.mch. Sub-range RR must be declared. It is not possible to write AA : j . . k → TT. The reason is that this can cause problems when translating to HDL. Constants may be declared in a type machine and must have an integer value:

CC = n ,

where n is an integer number. Enumerations may be declared in SETS-clauses:

EE = {K1, K2, …} .

Enumerations cannot be used as array index. This cannot be represented in all HDL target languages.

7. SUBSTITUTION LANGUAGE

The substitution language of BHDL, a subset of the substitution language of the B-language, does not contain a while-loop that is part of the implementable subset of the B language for software. BHDL-machines are expected to be cycle-accurate. So while-loops have to be encoded using state machines. To commence a development on a more abstract level Event-B can be used. There is no translation from Event-B but a path to BHDL suing a simple technique called event recomposition. The substitution language of BHDL comprises assignment, simultaneous substitution, sequential substitution, and conditional substitution. Arrays can only be assigned as a whole, i.e. assignments of the form x(k) := E are not possible. The reason is that tracking intermediary signals created in the translation would be complex and error-prone, whereas the price of the incurred restriction in practice is very low. We use lambda expressions to represent array values. So an array assignment has the form: $x := \lambda.(k \in K \mid E) \cup \lambda k.(k \in L \mid F) \cup \ldots$. The right-hand side may contain any number of lambda-expressions. The elements of the BHDL substitution language are interpreted in terms of hardware. Sequential substitution corresponds to causality, simultaneous substitution to concurrency, and conditional substitution to multiplexing. Registers are inferred from variables depending on their use. As a rule of thumb, a variable that is read before it is written represents a register. The BHDL substitution language consists of the following constructs:

x := E ,	(assignment)
x , y := E, F ,	(multiple assignment)
S || T ,	(simultaneous substitution)
S ; T ,	(sequential substitution)
IF P THEN S ELSE T END ,	(conditional substitution)
IF P THEN S END ,	
VAR x IN S END ,	(local variable substitution)
$x := \lambda.(k \in K \mid E) \cup \lambda k.(k \in L \mid F) \cup \ldots$	(multiple array assignment)

Their semantics is standard as described e.g. in the B-Book[1]. The last construct is used to model parallel hardware designs. Note that in multiple array assignment, the domain of the expression on the right hand side must equal the domain of variable x because x must be a total function (see preceding section). This is enforced by the B-language (and has to be proven).

8. THE TRANSLATION

The translation is based on the pre-post-predicate-transformation prdx described in [1] and the frame-based variant proposed in [2]. We extend the frames approach to read-frames and write-frames, which allows us to infer wires and registers from a circuit description. They are also used to verify that input and output variables are used properly. They are defined by

$read(S) =_{def}$ {variables read by substitution S} ,

$write(S) =_{def}$ {variables written by substitution S}.

Let *var*(*M*) denote the variables of a machine, *in*(*M*) the input variables, and *out*(*M*) the output variables. The following must hold for a machine *M* with operation *S*:

$in(M) \subseteq read(S) \wedge$
$in(M) \cap write(S) = \varnothing \wedge$
$out(M) \subseteq write(S) \wedge$
$out(M) \cap read(S) = \varnothing.$

Let *init*(*M*) denote the variables that initialized deterministically in machine *M*. Using *var*(*M*) and *init*(*M*) we define how registers and wires are inferred:

- The variables $read(S) \cap write(S)$ are translated to registers.
- The variables *init*(*M*) are translated to registers.
- All other variables are translated to wires.

Note that necessarily $read(S) \cap write(S) \subseteq var(M)$.

We show the principles of the translation by way of VHDL. Translation to SystemC, for instance, is similar. Based on the clauses INPUTS and OUTPUTS, an entity declaration is generated. A clock and a reset signal are added. The architecture body contains the block $\mathrm{tr}_y(S)$, where *S* is the operation of the BHDL-implementation and the processes describing the registers. And *y* denotes the complete list of the variables as stated in the VARIABLES and OUTPUTS clauses of the implementation. The definition of $\mathrm{tr}_z(S)$ follows. The substitutions are translated into a block of concurrent statements as described by translation $\mathrm{tr}_z(S)$:

$\mathrm{tr}_z(x := E) =$
$\quad x' \Leftarrow E;\ z' \Leftarrow z\,;$
tr_z (BEGIN *S* END) =
block begin $\mathrm{tr}_z(S)$ end block;
$\mathrm{tr}_z(S \parallel T) =$
block begin
$\quad\quad \mathrm{tr}_{z\text{-}write(T)}(S)$
$\quad\quad \mathrm{tr}_{z\text{-}write(S)}(T)$
end block;
$\mathrm{tr}_z\ (S\,;\,T) =$
block signal z'' : $type_z$; begin
$\quad\quad [z' := z'']\ \mathrm{tr}_z(S)$
$\quad\quad [z := z'']\ \mathrm{tr}_z(T)$
end block;
tr_z(VAR *x* IN *S* END) =
block signal x' : $type_x$; begin $\mathrm{tr}_{z\cup\{x\}}(S)$ end;
tr_z(IF *B* THEN *S* ELSE *T* END) =
block
$\quad\quad$ signal *gd* : *BOOLEAN*;
$\quad\quad$ signal z_t, z_e : $type_z$;

begin

$gd \Leftarrow B$;

$[z' := z]$ $\text{tr}_z(S)$

$[z' := z]$ $\text{tr}_z(T)$

$z' \Leftarrow z_t$ when gd else z_e ;

end;

The statement $x1, x2 := E1,E2$ is translated as $x1 := E2 \parallel x2 := E2$. The letter x denotes a variable, or a list of variables in which case an associated signal assignment statement should be read as multiple concurrent signal assignments. By E we denote any simple qualifier-free expression, or a union of lambda-expressions. The letters S and T denote substitutions.

The following names must not occur anywhere in a BHDL-implementation:

clock reserved for the clock signal introduced during the translation,
reset reserved for the reset signal.

The translator produces an internal intermediate hardware description which is subsequently translated into different target languages. Formal correctness of the first translation has been established. In the second step, it has been verified that the used language constructs in the target languages match the semantics of the intermediate language.

The translator produces an architecture consisting of three principle kinds of components:

COMB a combinatorial circuit,
REG some registers, and
CONN some connections.

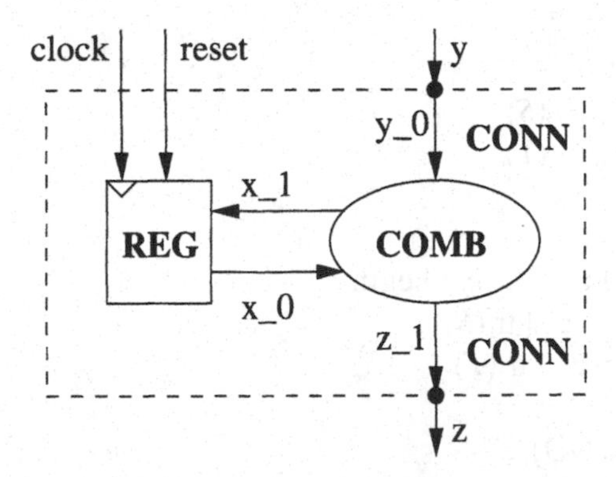

Figure 7-2. Generic schema of generated circuit

The combinatorial circuit **COMB** corresponds to the behavior specified by the operation of a BHDL-implementation. The set of registers **REG** corresponds to a subset of the variables of the implementation. The combinatorial circuit is connected to the inputs and outputs of the design by connections **CONN**.

9. EXAMPLES

Register : A register with reset is inferred for the following machine *REG_R*. If the initialization were non-deterministic initialization no reset would be generated. Initialization of machines is considered to create a defined state. This corresponds to a reset in hardware systems.

IMPLEMENTATION
    REG_R
INPUTS
    DIN
OUTPUTS
    DOUT
VARIABLES
    QQ
INVARIANT
    $DIN \in INT16 \wedge DOUT \in INT16 \wedge QQ \in INT16$
INITIALISATION
    $QQ := 0$
OPERATION
    $DOUT, QQ := QQ, DIN$
END

This can be represented as a schema:

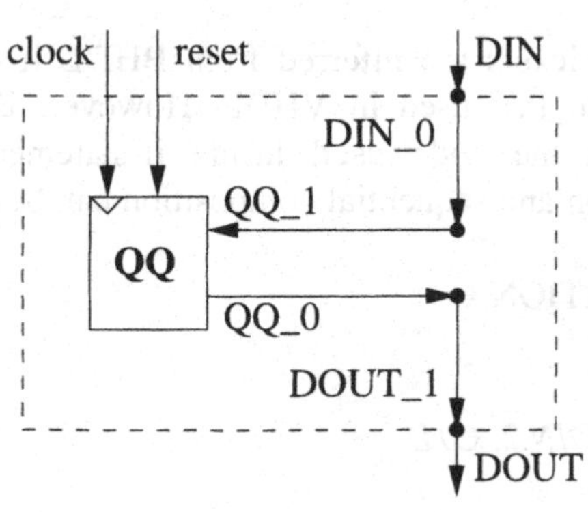

Figure 7-3. Register with reset

or in VHDL:

```
use BHDL.all;
entity REG_R is port (
        clock : in BIT;
        reset : in BIT;
        DIN : in INT16;
        DOUT : out INT16);
end;
architecture VHDL of REG_R is
signal DIN_0 : INT16;
signal DOUT_1 : INT16;
signal QQ_0 : INT16;
signal QQ_1 : INT16;
begin
process (clock, reset) begin
        -- initialisation is translated into an asynchronous reset
        if reset = '1' then
                QQ_0 <= 0;
        elsif clock'EVENT and clock='1' then
                QQ_0 <= QQ_1;
        end if;
end process;
COMB : block begin
        DOUT_1 <= QQ_0;
        QQ_1 <= DIN_0;
        DIN_0 <= DIN;
        DOUT <= DOUT_1;
end block;
end;
```

Multiplexer: Multiplexers are inferred from BHDL if-statements. The mechanism is similar to that used in VHDL. However, BHDL does not impose restrictions on the code used inside if-statements. Especially, simultaneous substitution and sequential composition can be mixed freely as everywhere else.

IMPLEMENTATION
    MUX
INPUTS
    DIN 1, *DIN* 2, *CTL*
OUTPUTS
    DOUT
INVARIANT
    $DIN1 \in INT16 \wedge DIN2 \in INT16 \wedge DOUT \in INT16 \wedge CTL \in BOOL$

```
OPERATION
        IF CTL = TRUE THEN
                DOUT := DIN 1
        ELSE
                DOUT := DIN 2
        END
END
```

The schema corresponding to the BHDL description is below:

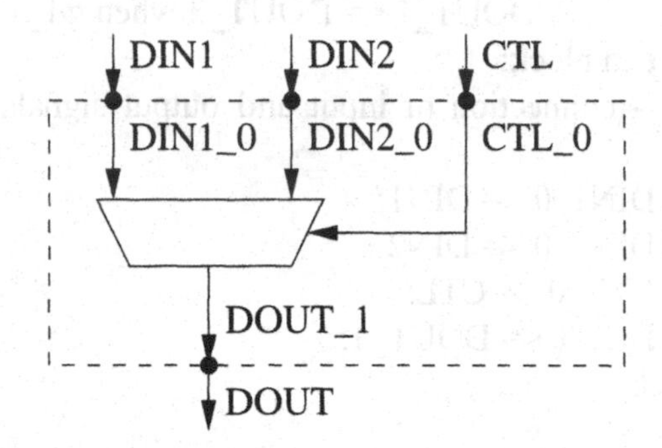

Figure 7-4. Multiplexer

The corresponding VHDL code is presented next. Usually the VHDL code is about 2 to 4 times as large as the corresponding code in BHDL.

```
use BHDL.all;
entity MUX is port (
        clock : in BIT;
        reset : in BIT;
        -- input and output signals as declared in BHDL
        DIN1 : in INT16;
        DIN2 : in INT16;
        CTL : in INT16;
        DOUT : out INT16);
end;
architecture VHDL of MUX is
signal DIN1_0 : INT16;
signal DIN2_0 : INT16;
signal CTL_0 : INT16;
signal DOUT_1 : INT16;
begin
-- no registers are generated
COMB : block begin
        CD_2 : block
```

```
        signal DOUT_3 : INT16; -- output signal of if-branch
        signal DOUT_5 : INT16; -- output signal of else-branch
        signal gd_3 : BOOL; -- guard condition of if-statement
        begin
                -- combinatorial computations
                gd_3 <= (CTL_0=true);
                DOUT_3 <= DIN1_0; -- computation of if-branch
                DOUT_5 <= DIN2_0; -- computation of else-branch
                -- this is the actual multiplexer in VHDL
                DOUT_1 <= DOUT_3 when gd_3 else DOUT_5;
        end block;
        -- connection of input and output signals to combinatorial
    circuit
        DIN1_0 <= DIN1;
        DIN2_0 <= DIN2;
        CTL_0 <= CTL;
        DOUT <= DOUT_1;
    end block;
    end;
```

10. REFERENCES

[1] Jean-Raymond Abrial. *The B-Book – Assigning programs to meanings*. Cambridge University Press, 1996.

[2] Steve Dunne. A Theory of Generalised Substitutions. In D. Bert, J. P. Bowen, M. C. Henson, and K. Robinson, editors, *ZB 2002: Formal Specification and Development in Z and B*, volume 2272 of *LNCS*, pages 270–290, 2002.

Chapter 8

TOWARDS A CONCEPTUAL FRAMEWORK FOR UML TO HARDWARE DESCRIPTION LANGUAGE MAPPINGS

Michele Marchetti, Ian Oliver

Nokia Research Center
Itämerenkatu 11-13
Helsinki,Finland

michele.marchetti@nokia.com, ian.oliver@nokia.com

1. INTRODUCTION

In this chapter we discuss a conceptual framework for the basis of a UML profile for hardware/software co-design and the meta-models of the target languages such as SystemC used in this domain.

UML 1.4/1.5[1] (Uml, 2002b, Uml, 2002a) is the defacto standard language for describing models with object-oriented concepts. The UML contains a meta-model written using the Meta-Object Facility language (MOF) to express the relationship between concepts it embodies.

Different domains and needs are supported by the UML profiling mechanisms which allows one to extend the core UML meta model and tailor the *deliberately weak* UML semantics and notation. Examples of this can be seen with the UML-RT and the Nokia Copenhagen UML profile (Oliver, 2002c).

The move to use UML for software-hardware co-design necessitates the incorporation of the concepts from these domains into the UML (Barros and López, 1999) to produce a UML-for-Hardware profile. Many of the concepts in this

J. Mermet (ed.), UML-B Specification for Proven Embedded System Design, 121–134.
© 2004 *Kluwer Academic Publishers. Printed in the Netherlands.*

field are embodied in the already existing hardware specification languages such as SystemC and VHDL[2].

In addition to the incorporation of these concepts into the UML we also have the Model Driven Architecture (MDA) driven by the OMG which has the idea of moving between platform independent and platform specific models via MDA mappings which map between concepts in the two levels of modeling abstraction. Central to this idea are the mappings from one set of UML profile's meta-models to another and how the modelling language changes or is extended as the modelling proceeds from abstract, platform independet to concrete, platform specific models.

In this paper we first present a brief review of mapping technologies from UML to hardware specification languages. We then present the SystemC and VHDL language meta-models and describe the conceptual mappings between a more abstract view of these languages. This more abstract view then provides us with a foundation of a conceptual framework for the development of a hardware-software-co-design profile for UML. We discuss then how one may use this profile and conclude with a discussion of future work.

2. RELATED WORK

Much previous work in the area of mapping UML to hardware specification languages has been done. However much of this work suffers from being too specific to one domain and also based upon a syntactical mapping between the UML language and the target language rather than more a conceptual, semantic based mapping we are proposing.

One of the earliest works in this area was made by Cheng et.al. regarding the mapping of UML and OMT to VHDL (McUmber and Cheng, 1999). This work concentrated on mapping the OMT class diagrams to the VHDL entity concept and the OMT state diagrams to the VHDL architecture concept through the application of a set of rules to the OMT models.

This method produced behavioural VHDL which is suitable for simulation but unsuitable for hardware synthesis which requires RTL VHDL.

Of interest is the UML to SpecC mapping known as YES-UML (Diaz-Herrera, 2001) which seeks to map to a more state chart based formalism. The paper does outline a set of stereotypes that are relevant in hardware modeling. It also defines the mapping between SpecC and UML in such as way that the two languages become isomorphic or homogenous in nature.

In (ao M Fernandes et al., 2000) a full methodology for developing hardware based embedded systems using UML with subsequent mapping to C and VHDL is described. This paper differs from the rest in that one is required to adhere to this particular methodology rather than use a mapping that could be applied in any industrial methodology such as Octopus (Awad et al., 1996) or ROOM (Selic et al., 1994).

3. META-MODELS OF HARDWARE SPECIFICATION LANGUAGES

In this section we describe the meta-models of our chosen target languages, ie: SystemC and VHDL. We then factor these meta-models in such a way that one can then build a more abstract conceptual representation or in MDA terms a more platform independent model which can be used to form the basis for a UML profile. The aim of this is to understand what information is necessary in order to construct a domain specific language suitable for use as a UML profile.

We must be careful to develop our abstract conceptual model in such a way that we do not exclude other hardware specification languages, eg: Verilog that we are not considering here. We must also not reinvent concepts that have already been defined in other relevant UML profiles, specifically UML-RT (uml, 2002). The aim of this is to create a generic language (or terms) that allow hardware engineers to model their designs without becomming platform or language specific.

The meta-models are constructed from the target languages by extracting both the grammar (syntax) of the language and the behavioural (semantic) components then using this information to build the meta-model (Wirfs-Brock et al., 1990). This means that we try to avoid a representation of the abstract syntax and concentrating on how the elements of the language behave together.

3.1 SystemC

SystemC is a combined language for hardware definition and test bench design. SystemC can also be used for systems' design (Grotker et al., 2002). In figure

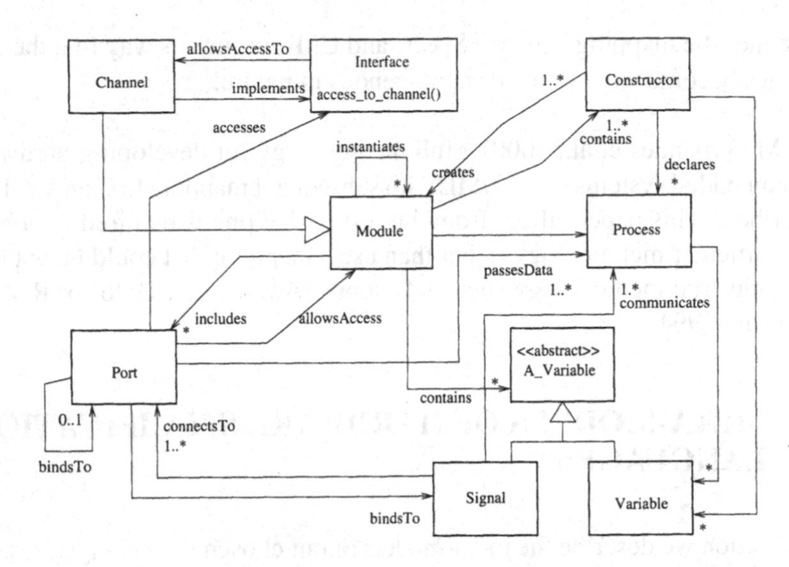

Figure 1.1. SystemC Meta-Model

1.1 we can see our meta-model for the SystemC language that outlines the major concepts in the language.

The main concept is the 'module', which is performs the function of being the main unit of design in SystemC. It is therefore related to almost all the others elements. As the relationship to itself shows, it might also be used to instantiate other modules.

A module might contain a number of 'abstract variables'. This concept doesn't belong to SystemC language implementation, but we have found it useful to add it in the conceptual description. This concept is used to abstract or factor the common concepts present in SystemC's 'signals' and 'variables'. Conceptually both of them can be used for similar purposes but have a different scope:

- signals are used to perform communications between components (for example, modules or processes)
- variables are used as local elements of a single component.

Another interesting component in SystemC is the 'channel' that from the conceptual point of view can bee seen a specialisation of module that has is designed to manage a data transfer medium.

An 'interface' is an object that is implemented by a channel and provides access to the physical medium transfer. This we have facilitated by the operation on interface - `access_to_channel()`. We can indeed generalize the functionality of this component from the conceptual point of view.

3.2 VHDL

VHDL is a dedicated hardware definition language. The language is designed to be used in two ways, firstly, to describe behaviour and then via refinement to a synthesisable level known as RTL VHDL. Interestingly VHDL began life as a documentation language rather than a specification language.

In figure 1.2 we present our VHDL meta-model.

Again as in the previous section we start by considering as most important elements: 'Entity' and 'Architecture'. These are often referred to as an or the Entity-Architecture pair in design documentation. In the meta-model note that we model this relationship as "an architecture realises and entity". We do not place any constraints on exactly how many architectures a particular entity may have.

An interesting group of relationships is between 'Entity', 'Architecture' and 'Component' elements. 'Components' are other modules (so again entity plus architecture pair) that are instantiated from an architecture (and might be placed into a library). By using OCL we have a better description of the relationship between these three classes:

context Component
inv: **self**.$entity.isImplementedBy$ $\rightarrow$ **includesAll** (**self**.$architecture$)

Another interesting aspect that is possible to see from the Meta-Model is about 'concurrent statement' element, that is seen like a kind of process (again we don't care about implementation details). Indeed we use the "normal process component" if it is instantiated inside the module, but if it is instantiated inside another process then we get a concurrent statement.

By relationships between 'signals' and 'processes' it is possible to see that a signal triggers evaluation of one or more processes and a process might assign values to a signal. Assignment operation is usually performed by one process but is possible to use resolution functions to allow to two different process to have a signal assignment statement to the same signal.

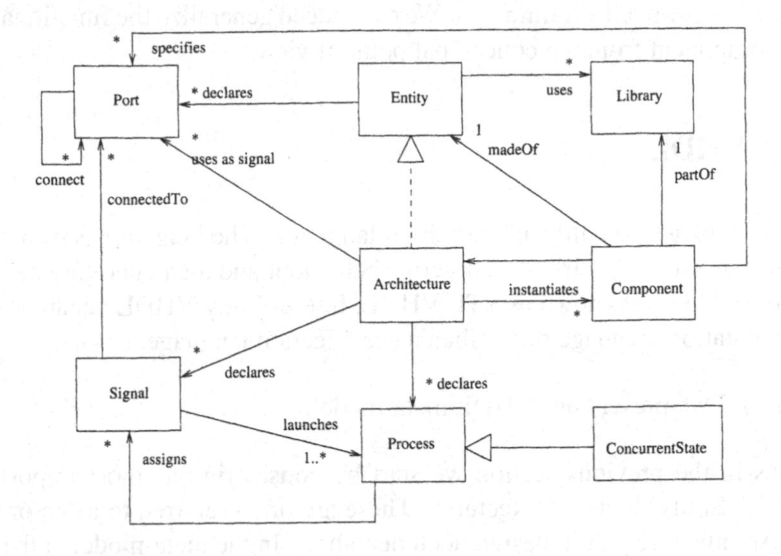

Figure 1.2. VHDL Meta-Model

3.3 Conceptual Mappings

By using the meta-models described in the previous section we could easily build a mapping between these two languages. However this mapping is not done from the semantic point of view but from a syntactic point of view. The problems with this is that a syntactic point of view does not take into consideration particular aspects of the semantics of the languages correctly into consideration.

An analogy to this is to consider a simpler, more common problem found in many tools. Consider a UML model where we have three classes A, B and C, where C inherits from both A and B. We can build purely syntactic mappings to Eiffel, C++ and Java. However we have the problem that we must decide in which order we inherit A and B, whether Eiffel's multiple inheritance semantics (Meyer, 1992) conflict with C++'s semantics and in the case of Java, how to perform something akin to multiple inheritance in a language that does not support this feature natively.

When building the conceptual mapping we must take into consideration that concepts in any pair of languages might not match exactly even though the concepts are considered to be the same. In table 3.3 we can see the conceptual mapping between VHDL and SystemC (VHDL concepts written horizontally and SystemC concepts vertically).

	Entity	Architecture	Port	Signal	Process	Variable
Module	X	X				
Process					X	
Port			X			
Signal				X		
Variable				X		X

Table 1.1. VHDL and SystemC Points of Commonality

In this table we can see the most interesting points of semantic or conceptual commanlity between the two languages, mainly the concept of a module in SystemC corresponds to that of an Entity-Architecture pair in VHDL and that a variable in SystemC can map to either a signal or variable in VHDL. One can also read these mappings inversely such that a VHDL signal is equivalent either a signal or variable in SystemC.

4. A FOUNDATION FOR A UNIFIED UML HARDWARE PROFILE

Of course just to develop meta-models of languages and extract their common parts is of no use unless it can be put into practice.

In the previous sections we have outlines the meta-models of the languages and the main points of semantic difference. However the differences are not just in structure but in behaviour, for example how does one map the synchronous and asynchronous semantics of RTL and Behavioural VHDL to SystemC and vice versa. Investigation into this area is proceeding.

If we are to follow the spirit of the OMG's Model Driven Architecture (MDA^{T}M) then these meta-models correspond to the MDA's platform specific concepts. We must now consider that we wish to model using a more platform independent approach and let the developer choose at a later point whether the model is to be implemented in either VHDL, SystemC or both. In order to facilitate this we require additional semantics to the UML that allows us to specify the major concepts directly in the model without being platform specific. By understanding the meta-models and their points of commonality and difference we can provide a basis for a *general set* of UML stereotypes that can be mapped cleanly down into the more platform specific VHDL and SystemC based models. This can be seen in figure 1.3 which shows the design flow. In other words what we are creating here is the foundation for building *domain specific languages*.

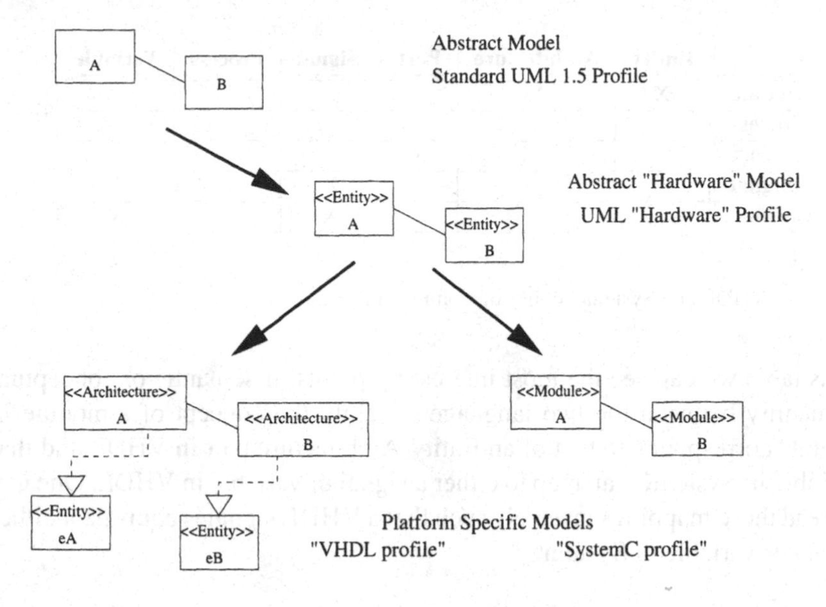

Figure 1.3. MDA Design Flow

We describe here now an example design flow in more detail demonstrating how these language mappings and conceptual language models can be used.

4.1 Model Driven Development

In this example we demonstrate how one may use MDA principles and the top-down modelling approach to develop a system. We consider here a highly simplified digital signal processor (DSP) architecture[3].

The first level of modelling is a highly platform independent model described using the core UML notation and semantics. Here we can see that a DSP contains simply a controller, protocols (eg: GSM, GPRS etc) and communication channels (which may be physical or logical). This is described using traditional object oriented modelling techniques.

MDA has the concept of a mapping which corresponds loosely to a step in some method which describes how that mapping should take place. A mapping may be automatic, semi-automatic or completely manual. Here we map the model to

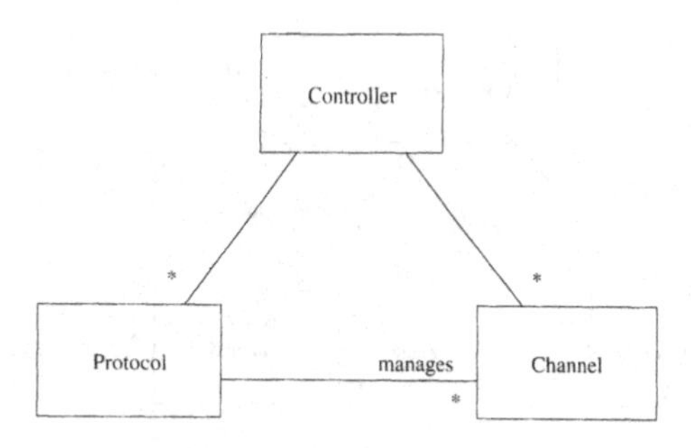

Figure 1.4. UML Model

a more platform specific model containing information about timing properties. In order to describe this we change the modelling language from core UML to UML-RT which contains the necessary modelling elements and language for describing these properties.

Using the UML-RT langauge we can add more information to the model regarding concurrency, process, threading etc. This may take the form of a number of MDA mappings to introduce this information. In addition we are making the model at each step more platform specific. Addition of this information corresponds to the bottom-up demands of the intended deployment.

There comes a point where we wish to decide upon the software-hardware partitioning of the model. Once this has taken place we now have two aspects to our model, one of which will be written using UML-RT and one which will be written using UML for Hardware. How this is noted in the models is not important here, we just state that the software-side models are contained in a UML package for convenience and simplicity here.

Because of the change of modelling language we now have the possibility to map the UML-RT modelling elements into the choices we have in UML-HW modelling elements. Our model may now appear as below.

Note how we could describe a channel class as an interface or entity and a selection of possible architectures. A similar idea could be applied to the controller class as well.

Mapping from this model into SystemC and VHDL is no longer a straightforward syntactic mapping from the UML model to the target language. A number of choices have to be made depending upon how we wish the target implementation to be made. For example the variables in a UML-HW model

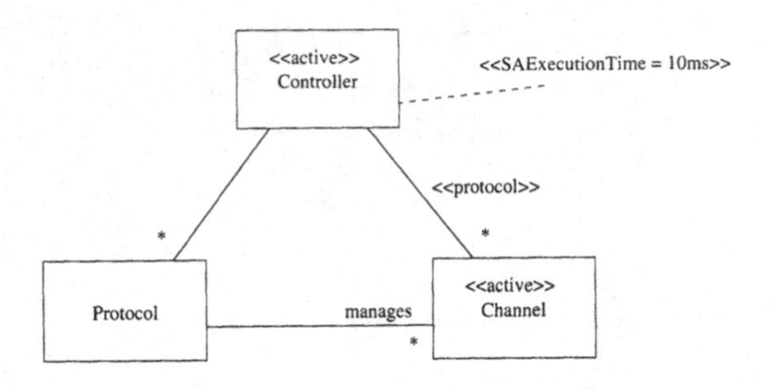

Figure 1.5. UML-RT Model

can map to either signals or variables in VHDL or just variables in SystemC. In the UML-HW model we may wish to force this by noting that certain variables do indeed become signals, in this case the mapping to SystemC would ignore this fact *or* generate possible extra code to capture this aspect.

Similarly we have the concept of VHDL's entity-architecture pairings in our UML-HW language. The mapping here from UML-HW to VHDL is reasonable straightforward but when mapping to SystemC we must choose which particular architecture to take. Actually this point is very clear to a VHDL user but not necessarily clear to someone who only reads UML or some other implementation language.

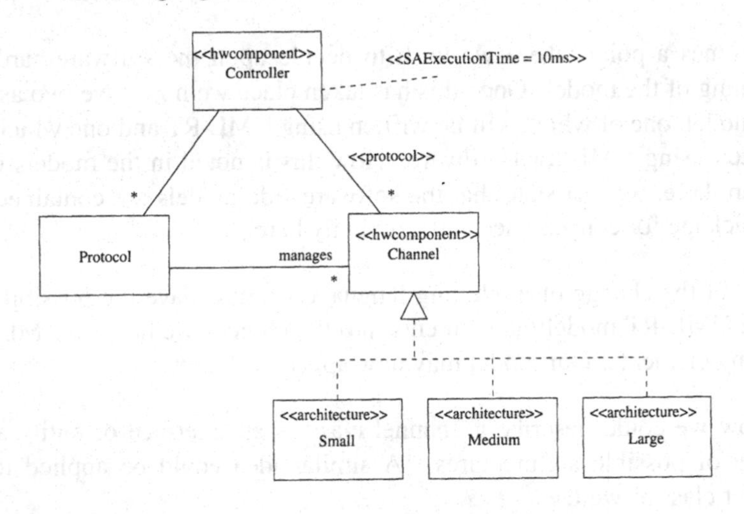

Figure 1.6. UML-HW Model

Here we can see three possible architectures for the channel hardware component. When mapping into a more specific level from here we can choose which particular architecture we use. If the architecture changes then this will be reflected down through the platform independent to platform specific mapping.

We are developing based upon our models UML profiles so that the designer can accurately map specific UML modelling elements to particular VHDL or SystemC elements. One can perform the mapping automatically, however, as we have seen with existing UML code generation tools the mapping is performed based upon syntax and not upon particular facets of the desired semantics. We prefer to give this control to the developer rather than some tool (Oliver, 2002a, Oliver, 2002b). Consider the situation where an abstract variable may be mapped to either a 'real' variable or a signal - how does a purely syntactic mapping choose?

The higher level platform independent profile should now take into consideration the common elements between the two languages (table 3.3). The main point in any model is the basic structure and this is represented by the entity-architecture pairings or by a module defintion. We have chosen at the more abstract level to use the stereotype $<< hwcomponent >>$ to represent these concepts and then only at the more platform specific levels does one worry about the implementation of this concept. The mappings are invisaged as shown in figure 1.7. Note that in figure 1.6 we expect that it may be useful to show some kinds of potential architecture, in which case one can easily see the correspondance with VHDL.

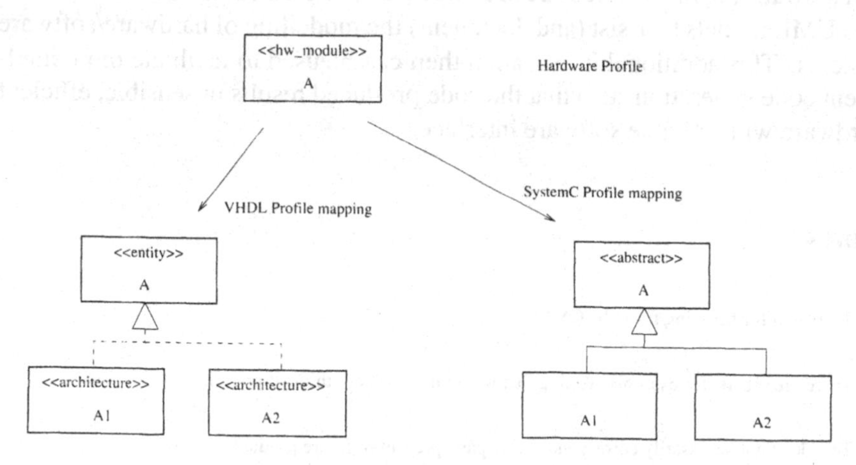

Figure 1.7. Structural Concept Mapping

The case with signals and variables is that at the abstract hardware level is that the choice would not be specified unless it was absolutely necessary. Only at the lower platform specific levels would the choice be made.

5. CONCLUSIONS AND FUTURE WORK

In this paper we have presented our VHDL and SystemC meta-models and shown the correspondence between the concepts in these models. We have also shown how one can use the UML/MOF to model these languages.

Using these models we have discussed that one can build a more generic form for languages in these classes and then construct mappings from this generic form into both (in our case) the more concrete or domain specific languages. In understanding the relationships between the concepts we can construct a more intelligent and valid mapping down into the concrete languages. This enables us to preserve the semantics we have in the model more correctly and consistently across any number of implementation or *platform specific* languages.

Much of this work is on-going and thus the future work is basically to refine and test the profile in larger models and ensure that the mappings do preserve the semantics such that a model that is mapped to both SystemC and VHDL does indeed behave in the same or similar (with known differences) manner. We must also consider in more detail how the use of a number of modelling languages throughout the development process is supported by suitable methodological practice.

What we aiming for is a structure that allows us to incrementally add information into UML models to assist (and document) the modelling of hardware/software systems. This additional information then can be used to facilitate more intelligent code generation such that the code produced results in sensible, efficient hardware with suitable software interfaces.

Notes

1. 2.0 is still forthcoming from the OMG

2. There are others, for example, Verilog, but we do not consider these at present

3. This does not necessarily correspond to any past, present or future product

References

(2002a). *OMG Unified Modelling Language Specification)*. Object Management Group, version 1.5 edition. OMG Document Number ad/02-09-02.

(2002b). *OMG Unified Modelling Language Specification (Action Semantics)*. Object Management Group, version 1.4 (final adopted specification) edition. OMG Document Number ad/02-01-09.

(2002). *Response to the OMG RFP for Schedulability, Performance and Time*. Object Management Group, revised submission edition.

ao M Fernandes, Jo Machado, Ricard J, and Santos, Henrique D (2000). Modeling industrial embedded systems with uml. In *In Proceedings of Eighth International Workshop on Hardware/Software Codesign CODES 2000, San Diego, California*.

Awad, Maher, Kuusela, Juha, and Ziegler, Jürgen (1996). *Object-Oriented Technology for Real-Time Systems. A Practical Approach Using OMT and Fusion*. Prentice-Hall.

Barros, Santiago Domínguez and López, Juan Carlos López (1999). Heterogeneous system design: A uml based approach. In *Proceedings of the 25th Euromicro Conference, Milan, Italy*, pages 386–389. IEEE Computer Society Press.

Diaz-Herrera, Jorge L. (2001). An isomorphic mapping for specc in uml. In *In Proceedings of OMER-2:Workshop on Object-oriented Modeling of Embedded Real-time Systems.*

Grotker, Thorsten, Liao, Stan, Martin, Grant, and Swan, Stuart (2002). *System Design with SystemC.* Kluwer Academic Publishers. 1402070721.

McUmber, William E. and Cheng, Betty H.C. (1999). Uml-based analysis of embedded sytems using a mapping to vhdl. In *High Assurance Software Engineering 1999.* IEEE.

Meyer, Bertrand (1992). *Eiffel, The Language.* Prentice Hall International.

Oliver, Ian (2002a). Experiences of model driven architecture in real-time embedded systems. In *Proceedings of FDL02, Marseille, France, Sept 2002.*

Oliver, Ian (2002b). Model driven embedded systems. In Lilius, Johan, Balarin, Felice, and Machado, Ricardo J., editors, *Proceedings of Third International Conference on Application of Concurrency to System Design ACSD2003, Guimarães, Portugal.* IEEE Computer Society.

Oliver, Ian (2002c). A uml profile for real-time system modelling with rate monotonic analysis. In *Proceedings of FDL02, Marseille, France, Sept 2002.*

Selic, Bran, Gullekson, Garth, and Ward, Paul T. (1994). *Real-Time Object-Oriented Modelling.* Wiley.

Wirfs-Brock, Rebecca, Wilkerson, Brian, and Wiener, Lauren (1990). *Designing Object-Oriented Software.* Prentice Hall.

Chapter 9

INTERFACE-BASED SYNTHESIS REFINEMENT IN B

T. Lecomte
ClearSy

1. INTRODUCTION

When specifying a system, the refinement process of the B method enables to refine the behaviour (inner part) and the interface (outer part) of a component, from its abstraction to its implementation, while verifying, by proof at each refinement step, that there is no contradiction between two successive refinement levels

For one abstract component, there are many suitable implement-able components. Suitability is ensured by proof. Thus, in a system specification,

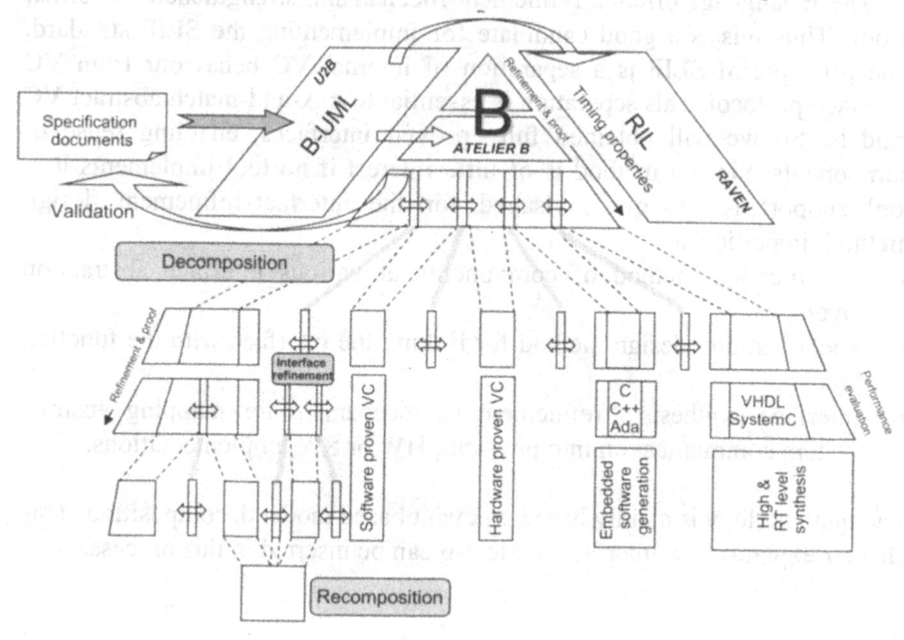

J. Mermet (ed.), UML-B Specification for Proven Embedded System Design, 135–154.
© 2004 *Kluwer Academic Publishers. Printed in the Netherlands.*

components can be kept abstract, which enables the designer to choose afterwards which component is to be used. In this case, the refinement process has to be adapted in order to converge to the targeted component. This approach obviously allows to combine legacy components, by reusing existing components, replacing some ones by up to date components, in a word by enabling the system designer to do the very work of an engineer but within the framework of a proof validated development process.

This document aims at providing guidelines concerning the use of B at system level to specify interface-based components, refine the interface viewed as a component itself and to reuse existing components.

2. DESCRIPTION

Our objective is to provide a design method for the system-level interface-based design as defined in [1], and is driven but the needs for:

- Reuse at the system level of system houses know-how rather than implementation reuse,
- Rise the abstraction level for design and integration,
- Design of complex systems from IP blocks provided at various abstraction levels in different formalisms/notations.

The B language offers a refinement mechanism, strengthened by formal proof. Thus this is a good candidate for implementing the SLIF standard. The principle of SLIF is a separation of internal VC behaviour from VC interface protocol. This separation is essential to mix-and-match abstract VC models. So we will obtained fully proved interfaces, enabling reuse of components. Since a method is of little interest if no tool implements it, a tool support is strongly envisaged for the interface-refinement design method; in particular:

- Specification method of components at various interface abstraction layers,
- Specification / design method for linking the interface with the function block,
- Interface synthesis / refinement (a mechanism for mapping abstract system communication into physical, HW or SW, implementations.

The methodology is mainly based on event based model decomposition. This chapter explains how interface modeling can be inserted in this process,

3. METHOD

In a B system-level model, the interface definition is performed while distributing functions over sub systems. More precisely, the process is composed of four successive steps:

1) Writing of the initial model,
2) Preliminary model decomposition,
3) Definition of the relationships between sub-systems
4) Interface description

3.1 Writing the initial model

By initial model, we mean any model composed of one specification component and zero or more refinement components. All these components are refined by only one other component, except the last one (the most concrete), which has to be decomposed. This model is called a refinement column.

This model may have been previously obtained by decomposing an other system or sub-system model.

Such a model is event-based and describes a closed system.

3.2 Preliminary model decomposition

By applying the "divide and conquer" strategy, the designer has to partition the set of events resulting from its last refined component. He must:

- determine the number of required sub-systems,
- name the sub-systems,
- allocate events to sub-systems.

Each event should be allocated to one and only one sub-system.

3.3 Relationships between sub-systems

Relationships are determined by the variables distribution. The objective of this step is to determine how variables are to be shared among sub-systems (location, access type).

The following tasks have to be performed:

- Listing variables,
- Finding dependencies among sub-systems.

A variable can be accessed according to the following modes: R (Read), W (Write), RW (Read/Write), NA (NoAccess).

Variables can be sorted out as described below:

Condition	Classification
A variable is accessed in R mode once or more, never modified (W).	Allocate to one sub-system with Read access interface
A variable is accessed in R mode once or more, modified (W) by only one sub-system	Sub-system local variable. access interface.
A variable is accessed in R mode once or more, modified (W) by more than one sub-system	Allocate to one sub-system with Read / Write access interface and synchronization
A variable is never accessed in R mode, modified (W) by one sub-system	Sub-system local variable.
A variable is never accessed in R mode, modified (W) by more than one sub-system	Allocate to one sub-system with Write access interface and synchronization
A variable is never accessed (NA).	Dead variable

Let us now consider the following model described below.

```
  SYSTEM S0
SETS DATA
VARIABLES
   d1, d2, b1, b2
INVARIANT
   d1: DATA & d2: DATA
INITIALISATION
   d1 :: DATA || d2 :: DATA
EVENTS
   evt1 = SELECT b1=TRUE THEN
      d1 :: DATA  || b2 := TRUE
   END;
   evt2 =  SELECT b2=TRUE THEN
      b1 := FALSE
   END;
   evt3 = SELECT b1=FALSE & b2=TRUE THEN
      b1=TRUE || d2 := d1
   END;
   evt4 = SELECT b1= b2 THEN
      d2 :: DATA
   END;
END
```

Access mode	d1	d2	b1	b2
evt1	W	NA	R	W
evt2	NA	NA	W	R
evt3	R	W	RW	R
evt4	NA	W	R	R

If we consider grouping evt1 and evt2 in sub-system SS1 and evt3 and evt4 in SS2, we obtain the following table:

Access mode	d1	d2	b1	b2
SS1	W	NA	RW	RW
SS2	RW	RW	RW	R

In this case, d1, b1 and b2 are shared by SS1 and SS2. D2 is local to SS2. We can draw the following diagram:

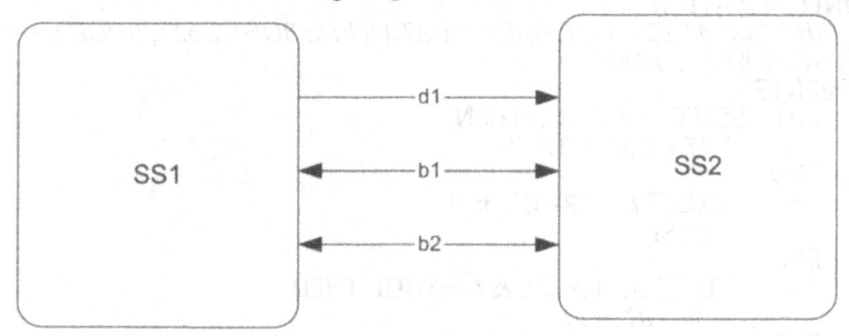

d1, b1 and b2 are used by both sub-systems, while d2 is local to SS2. In order to be able to separate SS1 and SS2, "shared" variables need to be allocated to one and only sub-system and the necessary interface protocol to be defined.

Let us consider that d1 and b1 are located in SS1 while b2 is located in SS2. So we need to create and manage copies of these variables in all the sub-systems where they are defined.

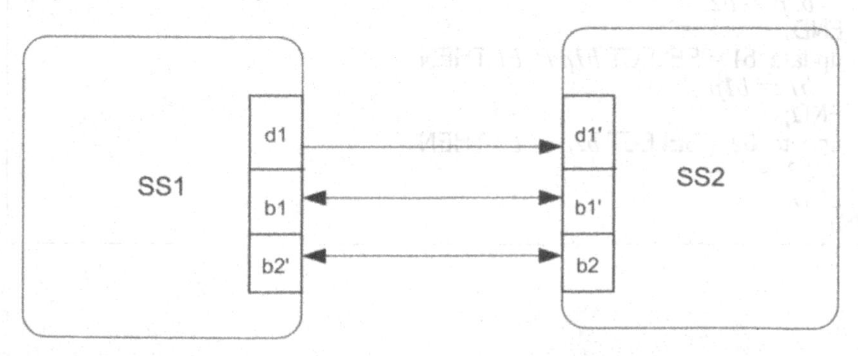

In this case, we need to define the way copies are updated by adding extra events. For example, for d1':

```
   update_d1 = SELECT d1' /= d1 THEN
     d1' := d1
END
```

We finally obtain the following model:

```
   REFINEMENT S1
REFINEMENT S0
VARIABLES
   d1, d2, b1, b2, d1p, b1p, b2p
INVARIANT
   d1: DATA & d2: DATA & d1p : DATA & b1 : BOOL & b2 : DATA & b1p
: BOOL & b2p : BOOL
INITIALISATION
   d1 :: DATA || d2 :: DATA || d1p :: DATA || b1 :: BOOL || b2 :: BOOL || b1p
:: BOOL || b2p :: BOOL
EVENTS
   evt1 = SELECT b1=TRUE THEN
      d1 :: DATA  || b2 := TRUE
   END;
   evt2 =  SELECT b2=TRUE THEN
      b1 := FALSE
   END;
   evt3 = SELECT b1=FALSE & b2=TRUE THEN
      b1=TRUE || d2 := d1
   END;
   evt4 = SELECT b1= b2 THEN
      d2 :: DATA
   END;
   update_d1p = SELECT d1p /= d1 THEN
      d1p := d1
   END;
   update_b1p = SELECT b1p /= b1 THEN
      b1p := b1
   END;
   update_b2p = SELECT b2p /= b2 THEN
      b2p := b2
   END;
   update_b1 = SELECT b1p /= b1 THEN
      b1 := b1p
   END;
   update_b2 = SELECT b2p /= b2 THEN
      b2 := b2p
   END
END
```

Updates are not ordered, as in the previous diagram, data flows are not precisely defined (no conditions). So events are allowed to modify replicated variables.

We can also introduce a control flow concerning d1:

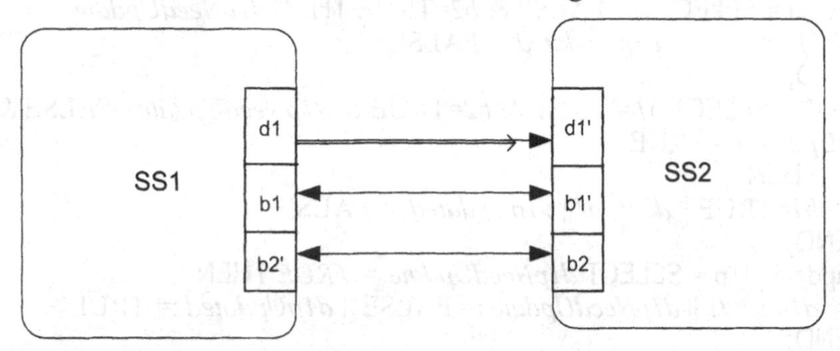

In that case, for each update of d1, SS2 is aware of that update. A new variable d1Modified (boolean) is added and some events need to be modified:

```
    evt1 = SELECT b1=TRUE THEN
    d1 :: DATA || b2 := TRUE || d1Modified := TRUE
END;
evt3 = SELECT b1=FALSE & b2=TRUE & d1Modified=FALSE THEN
    b1=TRUE || d2 := d1
END;
update_d1p = SELECT d1Modified = TRUE THEN
    d1p := d1 || d1Modified := FALSE
END;
```

Event evt3 is prevented to be fired when d1 has been modified while d1p has not been made equal to d1. Event evt3 is enabled as soon as event update_d1p has been triggered.

Now, if we modify the control flow on d1 according to the following diagram:

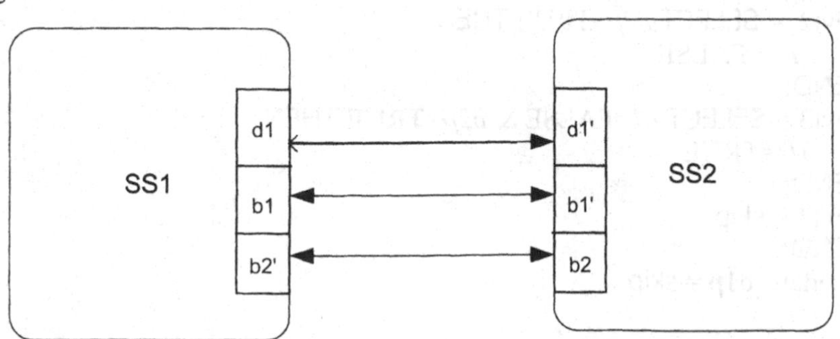

Some events need to be modified:

```
   evt1 = SELECT b1=TRUE THEN
     d1 :: DATA  || b2 := TRUE
   END;
   evt3a = SELECT b1=FALSE & b2=TRUE THEN d1pNeedUpdate
:=TRUE ||          d1pUpdated := FALSE
   END;
   evt3 = SELECT b1=FALSE & b2=TRUE & d1pNeedUpdate =FALSE &
d1pUpdated = TRUE
     THEN
     b1=TRUE || d2 := d1|| d1pUpdated := FALSE
   END;
   update_d1p = SELECT d1pNeedUpdate = TRUE THEN
     d1p := d1 || d1pNeedUpdate := FALSE || d1pUpdated := TRUE
   END;
```

A new event evt3a is added, as well as variables *d1pNeedUpdate* and & *d1pUpdated* (both boolean). Then, event evt3 is prevented to be fired as long as events evt3a and update_d1p have not been fired.

According to [2], our model can be decomposed into three sub-systems: SS1, SS2 and the interface between SS1 and SS2.

We obtain the following models:

```
 SYSTEM SS1
VARIABLES
   d1, b1, d1p, b2p
INVARIANT
   d1: DATA  & b1 : BOOL & b2p : BOOL
INITIALISATION
   d1 :: DATA || b1 :: BOOL || b2p :: BOOL
EVENTS
   evt1 = SELECT b1=TRUE THEN
     d1 :: DATA  || b2p := TRUE
   END;
   evt2 =  SELECT b2p=TRUE THEN
     b1 := FALSE
   END;
   evt3 = SELECT b1=FALSE & b2p=TRUE THEN
     b1:=TRUE
   END;
   evt4 = skip
   END;
   update_d1p = skip
```

```
  END;
  update_b1p = skip
  END;
  update_b2p = BEGIN
    b2p :: BOOL
  END;
  update_b1 = BEGIN
    b1 :: BOOL
  END;
  update_b2 = BEGIN
    b2 :: BOOL
  END;
END
```

```
   SYSTEM SS2
VARIABLES
  d2, b2, d1p, b1p
INVARIANT
  d2: DATA & d1p : DATA & b2 : DATA & b1p : BOOL
INITIALISATION
  d2 :: DATA || d1p :: DATA || b2 :: BOOL || b1p ::
EVENTS
  evt1 = skip
  END;
  evt2 =  skip
  END;
  evt3 = SELECT b1p=FALSE & b2=TRUE THEN
    b1p=TRUE || d2 := d1
  END;
  evt4 = SELECT b1p= b2 THEN
    d2 :: DATA
  END;
  update_d1p = skip
  END;
  update_b1p = skip
  END;
  update_b2p = skip
  END;
  update_b1 = skip
  END;
  update_b2 = skip
  END;
END
```

```
SYSTEM I_SS1_SS2
VARIABLES
  d1, b1, b2, d1p, b1p, b2p
INVARIANT
  d1: DATA & d1p : DATA & b1 : BOOL & b2 : DATA & b1p : BOOL &
b2p : BOOL
INITIALISATION
  d1 :: DATA || d1p :: DATA || b1 :: BOOL || b2 :: BOOL || b1p :: BOOL ||
b2p :: BOOL
EVENTS
  evt1 = BEGIN
     d1 :: DATA  || b2 :: BOOL
  END;
  evt2 =  BEGIN
    b1 :: BOOL
  END;
  evt3 = BEGIN
    b1:: BOOL
  END;
  evt4 = skip
  END;
  update_d1p = SELECT d1p /= d1 THEN
    d1p := d1
  END;
  update_b1p = SELECT b1p /= b1 THEN
    b1p := b1
  END;
  update_b2p = SELECT b2p /= b2 THEN
    b2p := b2
  END;
  update_b1 = SELECT b1p /= b1 THEN
    b1 := b1p
  END;
  update_b2 = SELECT b2p /= b2 THEN
    b2 := b2p
  END;
END
```

In those models, as explained in [2], shared variables can't be refined any further, as they appear in many models. They can be considered as "sockets" that are to be kept all along the modeling. So b1, b2, d1, b1p, b2p, d1p should not be refined.

All those subsystems SS1, SS2 and I_SS1_SS2 are refined by S1.

4. CASE STUDY: A SAFE POWER CHECKER

4.1 Presentation

The system described in this case study is aimed to check whether the power going through a certain electrical wire is bound between two limit values that may depend upon some external conditions. So the system has mainly one Boolean output used to set an alarm when the power of the electrical wire is not correct.

At this point, the system seems quite simple, except that it is a safety device, so extra care should be taken to compute the output. Concerning safety, the system output may very well be "power not correct", even if the power is actually correct, but it should never be "power correct" if the power is actually incorrect. In fact, the "never" used previously means precisely a MTBF greater than 10^9 hours.

4.2 Modeling

A B model is used to formalize this system from the very abstract system specification to the design of the embedded software. Different techniques are used: event B, refinement, event splitting, event decomposition, UML State Charts.

The first level modeled here describes the system specification. At this point, the main concern is the safety requirements. As the abstraction level is high. This first draft just gives the operational specification of the system according to the following State Chart. The SPC system is described with the following states:

- SPC_TurnedOff:
- SPC_TurnedOn: when the SPC is turned on. This state is composed of the following sub-states:
- SPC_Init: the SPC initialization, after it has been turned on,
- SPC_Nominal: the SPC nominal functioning mode, coming after the initialization and when no failure has been detected,
- SPC_Failure: when a failure has been detected.

The safety requirements says that in all functioning modes except nominal, the system output should be incorrect: "Ext_Power_Ok = FALSE".

State Charts are used in this case study to complete the B model. By giving some graphical representations focusing on the system states and the transitions between these states, they help to structure the B model, which is only a linear text. This method can be qualified as "Design and Substance" since it combines state of the art semi-formal design (UML State Charts) and formal model bringing substance to it (event B).

State Charts are more or less always formalized the same way in event B. A list of all possible states at some level of a State Charts is formalized by an enumerated set that may be the Boolean set when there are only two states. Then state variables are used for each state level to describe in which particular state the system is.

For instance, the first state level of the SPC system describing whether the system is turned on or off is formalized with set BOOL and variable SPC_Power_Ok. The sub-state of the system being turned on is formalized with the enumerated set:

SPC_STATES = {SPC_Init, SPC_Nominal, SPC_Failure}

and with variable SPC_State. These state variables are used in the guards of

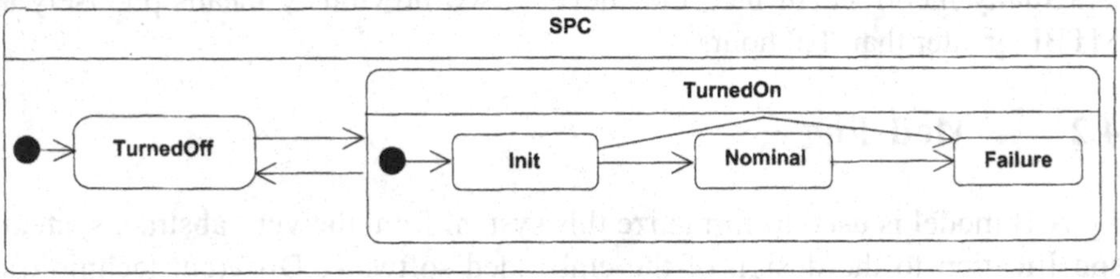

Figure 1: higher level state chart

B events and they are modified in the substitutions of B events. We try to deal with them at the beginning of guards and substitutions to help the reader of the B model distinguish between the UML structure and the rest of the B model.

4.2.1 System Design: General Level

In the first refinement of the SPC system, the major architecture decision is introduced. In order to be able to keep the safety requirement (10^{-9} failure/hour of functioning), redundancy is used. To do so, the system is build around two nearly identical sub-systems (see figure 2); each of them computes an output. The system output is the logical AND of the two outputs. In this way the power checked by the SPC is judged correct exactly when the two sub-systems judge it correct.

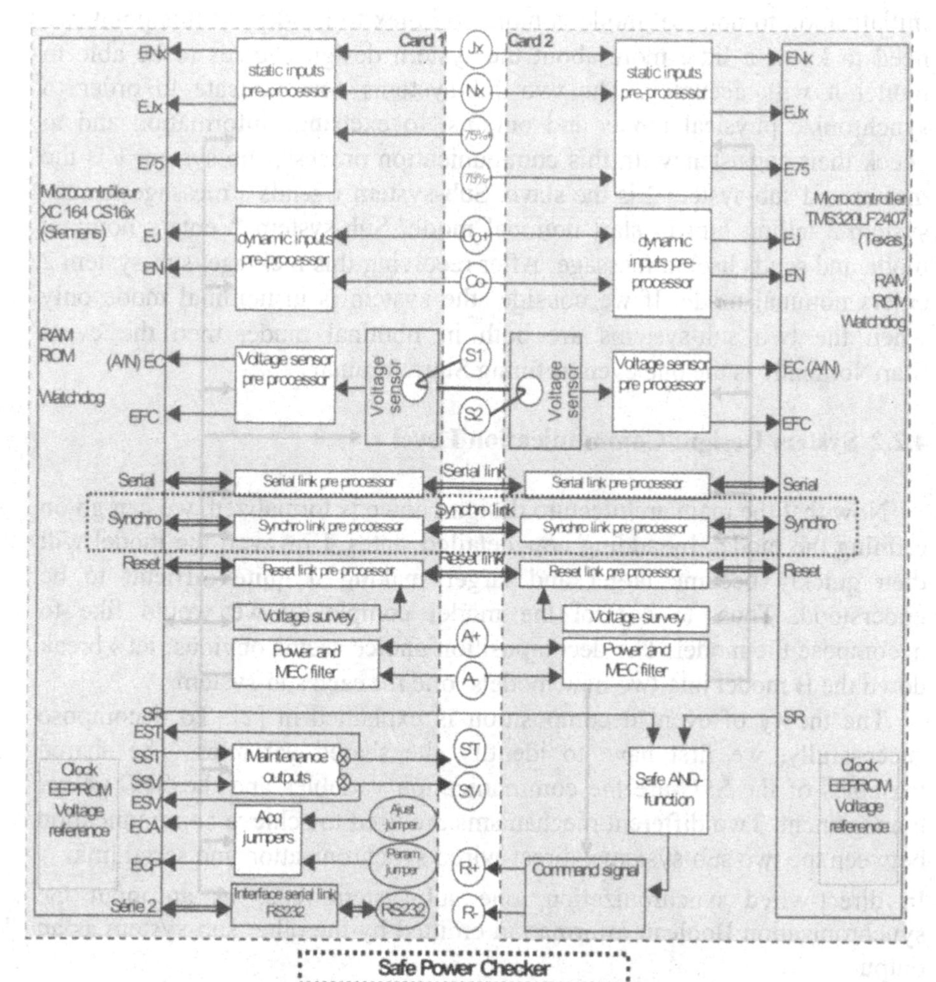

Figure 2: safe power checker design

So now we have two nearly identical State Charts instead of one, and we double variables and events from the previous component. The name of variables and events are post fixed by the number of the system they refer to (1 for sub-system 1 and 2 for sub-system 2).

In B, we also have to define precisely the relation between the current variables and the previous ones. This is done in the refinement gluing invariant. For instance, the system output is defined by:

Ext_Power_Ok = bool(Ext_Power_Ok1 = TRUE & Ext_Power_Ok2 = TRUE)

The refinement of the main functional modes also has to be defined. We consider that a failure has been detected in the whole system when a failure has been detected in at least one of the two sub-systems. The transition from initialization to nominal mode is more complex to handle. At this point, we need to know a little more about the system design choices to be able to model it with accuracy. The two sub-systems communicate in order to synchronize physical inputs and outputs, to exchange information and to check their consistency. In this communication process, sub-system 1 is the master and sub-system 2 is the slave. Sub-system 1 sends a message to sub-system 2 telling him to start nominal mode. Sub-system 2 enters nominal mode and sends back a message. After receiving this message, sub-system 2 enters nominal mode. If we consider the system is in nominal mode only when the two sub-systems are both in nominal mode, then the event StartNominal1 is the one event refining StartNominal.

4.2.2 System Design: Communication Level

Now that the main architecture design choice is formalized, we can go on refining the model, by adding new detailed states. However, the model will then quickly become larger and larger making it quite difficult to be understood. Thus, to control the model complexity we would like to decompose the model. The decomposition choice seems obvious: let's break down the B model into two new models, one for each sub-system.

The theory of event decomposition is explained in [2]. To decompose successfully, we first have to identify the shared variables. The shared variables of the SPC are the communication variables and the SPC power management. Two different mechanisms are used to achieve communication between the two sub-systems: direct-wired synchronization and serial link.

In direct-wired synchronization, one sub-system reads as an input the synchronization Boolean information emitted by the other sub-system as an output.

In serial link communication, one sub-system build a message, it sends a request to the distant sub-system and when they are both ready the message is sent and is eventually written in a buffer of the distant subsystem.

In synchronized communication, the Boolean synchronized state of each sub-system is a shared variable. And in serial link communication, the buffers where the messages are stored are the shared variables.

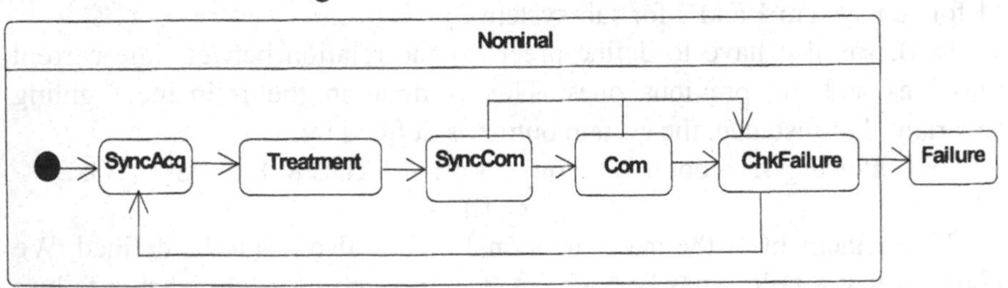

Figure 3: zoom in the nominal state

We have seen that in a decomposition process, shared variables will have to remain unchanged through the refinement of each sub-model. So here, we have to use the final concrete variables that will be used during coding. In fact, this is nothing more than defining the interface of the two sub-systems,.

Now we can build the next refinement level, by adding the communication variables, designing the new states modifying these variables, and adding the B events corresponding to the transitions between these new states. The new states are:

- SyncAcq: synchronization before starting acquisition,
- Treatment: treatments of the input data and computation of the output,
- SyncCom: synchronization before communication,
- Com: communication between the 2 sub-systems,
- ChkFailure: comparison of data of the two sub-systems and failure decision.

4.3 Refining interfaces

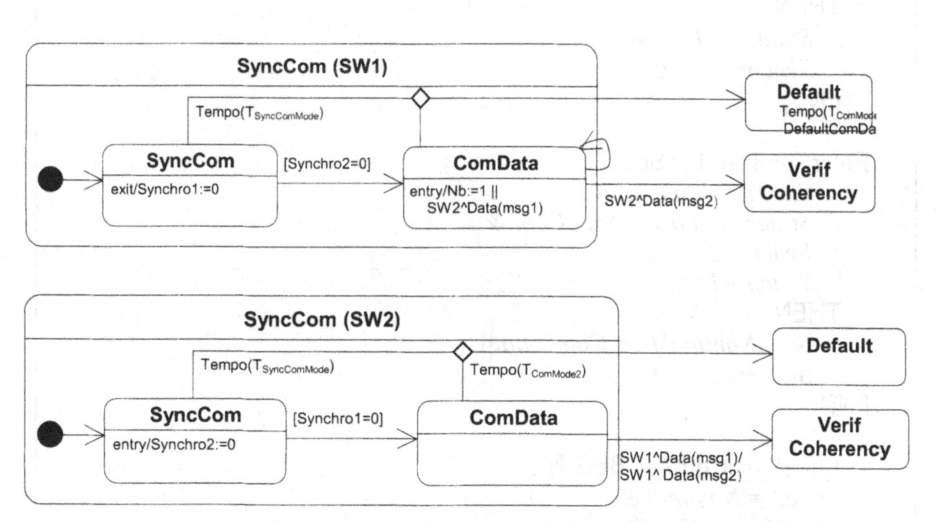

Figure 4: refinement of the synchronisation communication state

As the system design is based on a master/slave schema, the next step is to distinguish master and slave communication. As an example, interface refinement is applied the synchronisation communication state (SyncCom in figure 3). Two state charts (see figure 4) are elaborated for software 1 (master) and for software 2 (slave).

The synchronisation link is surrounded by a dotted line box in figure 2.

The event B model related to the synchronisation communication is listed below. The B model contains 3 events for the master and 3 events for the slave (enter, leave, timeout).

```
 REFINEMENT SPC_n
(...)
 EVENTS
(...)
 EnterSyncCom1 = BEGIN
   State1 = Nominal &
   StateNominal1 = Treatment
     THEN
     StateNominal1 := SyncCom
 END;

 TimeOutSyncCom1 = SELECT
   State1 = Nominal &
     StateNominal1 = SyncCom &
     Synchro2 = 1
   THEN
     State1 := Default ||
     Synchro1 := 0
   END;

 EnterComData1 = SELECT
     State1 = Nominal &
     StateNominal1 = SyncCom &
     Synchro2 = 0 &
     Synchro1 = 1
   THEN
     StateNominal1 := ComData ||
     Synchro1 := 0
 END;

   EnterSyncCom2 = BEGIN
   State2 = Nominal &
   StateNominal2 = Treatment
     THEN
     StateNominal2 := SyncCom
 END;

 TimeOutSyncCom2 = SELECT
   State2 = Nominal &
     StateNominal2 = SyncCom &
     Synchro1 = 1
```

```
    THEN
        State2 := Default ||
        Synchro2 := 0
    END;

  EnterComData2 = SELECT
        State2 = Nominal &
        StateNominal2 = SyncCom &
        Synchro1 = 0 &
        Synchro2 = 1
    THEN
        StateNominal2 := ComData ||
        Synchro2 := 0
  END
END
```

Variables *synchro1* and *synchro2* are shared and should be unshared in order to get clean interfaces: according to the schema described in §3, variable *synchro1* is allocated to the sub-system 1 (master, the one which embeds software 1) and *synchro2* is allocated to the sub-system2 (slave). New variables *synchro1'* and *synchro2'* are introduced.

New communications events are added to represent explicit data transfer between sub-system1 and subsystem2:

```
    updateSynchro1 = BEGIN
    Synchro1' ≠ Synchro1 &
    Synchro1'ToBeUpdated = TRUE
       THEN
       Synchro1' := Synchro1 ||
       Synchro1'ToBeUpdated := FALSE
  END;

    updateSynchro2 = BEGIN
    Synchro2' ≠ Synchro1 &
    Synchro2'ToBeUpdated = TRUE
       THEN
       Synchro2' := Synchro1 ||
       Synchro2'ToBeUpdated := FALSE
  END
```

The invariant should be extended with:

```
(Synchro1'ToBeUpdated = FALSE => Synchro1' = Synchro1) &
(Synchro2'ToBeUpdated = FALSE => Synchro2' = Synchro2)
```

and existing events should be rewritten as:

```
  TimeOutSyncCom1 = SELECT
    State1 = Nominal &
      StateNominal1 = SyncCom &
      Synchro2'ToBeUpdated = FALSE &
      Synchro2' = 1
   THEN
      State1 := Default ||
      Synchro1'ToBeUpdated := TRUE ||
      Synchro1 := 0
   END;
  EnterComData1 = SELECT
      State1 = Nominal &
      StateNominal1 = SyncCom &
      Synchro2'ToBeUpdated = FALSE &
      Synchro2' = 0 &
      Synchro1 = 1
   THEN
      StateNominal1 := ComData ||
      Synchro1'ToBeUpdated := TRUE ||
      Synchro1 := 0
  END;
  TimeOutSyncCom2 = SELECT
   State2 = Nominal &
      StateNominal2 = SyncCom &
      Synchro1'ToBeUpdated = FALSE &
      Synchro1' = 1
   THEN
      State2 := Default ||
      Synchro2'ToBeUpdated := TRUE ||
      Synchro2 := 0
   END;

  EnterComData2 = SELECT
      State2 = Nominal &
      StateNominal2 = SyncCom &
      Synchro1'ToBeUpdated = TRUE ||
      Synchro1' = 0 &
      Synchro2 = 1
   THEN
      StateNominal2 := ComData ||
      Synchro1'ToBeUpdated := TRUE ||
      Synchro2 := 0
  END
END
```

The next step [3] is to decompose the model into three sub-systems (sub-system1, sub-system2 and interface). All these steps are likely to be executed by a (semi) automatic tool.

The resulting interface component *I_SPC1_SPC2*, as any B component, is then suitable for stepwise refinement and decomposition.

```
SYSTEM I_SPC1_SPC2
VARIABLES
   Synchro1, Synchro1', Synchro2, Synchro2',
   Synchro1'ToBeUpdated, Synchro2'ToBeUpdated
INVARIANT
   Synchro1: DATA & Synchro1': DATA &
   Synchro2: DATA & Synchro2': DATA &
   Synchro1'ToBeUpdated: BOOL & Synchro2'ToBeUpdated: BOOL
(...)
EVENTS
(...)
updateSynchro1 = BEGIN
   Synchro1' ≠ Synchro1 &
   Synchro1'ToBeUpdated = TRUE
      THEN
      Synchro1' := Synchro1 ||
      Synchro1'ToBeUpdated := FALSE
END;

updateSynchro2 = BEGIN
   Synchro2' ≠ Synchro2 &
   Synchro2'ToBeUpdated = TRUE
      THEN
      Synchro2' := Synchro2 ||
      Synchro2'ToBeUpdated := FALSE
END;
TimeOutSyncCom1 = BEGIN Synchro2'ToBeUpdated :: BOOL END;
EnterComData1 = BEGIN Synchro1'ToBeUpdated :: BOOL END;
TimeOutSyncCom2 = BEGIN Synchro2'ToBeUpdated :: BOOL END;
EnterComData2 = BEGIN Synchro1'ToBeUpdated : BOOL END
```

To demonstrate the interface refinement process, let us consider inserting a synchronization enabler (for example, Synchro1 and Synchro2 evolve to high frequency, and we want our system to handle some of these values).

A new variable, *SynchroEnabler*, is introduced and represents the capability of taking into account new values of *Synchro1* and *Synchro2*. This variable, *SynchroEnabler*, may be refined further by a timer, The new model is given below, iwith modified events (updateSynchro1, updateSynchro2) and new enabling/disabling events (enableSynchro, disableSynchro):

```
REFINEMENT I_SPC1_SPC2_1
REFINES I_SPC1_SPC2
VARIABLES
  (...), SynchroEnabler
INVARIANT
  SynchroEnabler: BOOL
EVENTS
(...)
updateSynchro1 = BEGIN
  Synchro1' ≠ Synchro1 &
  Synchro1'ToBeUpdated = TRUE &
  SynchroEnabler = TRUE
    THEN
    Synchro1' := Synchro1 ||
    Synchro1'ToBeUpdated := FALSE ||
    SynchroEnabler := FALSE
END;
updateSynchro2 = BEGIN
  Synchro2' ≠ Synchro2 &
  Synchro2'ToBeUpdated = TRUE&
  SynchroEnabler = TRUE
    THEN
    Synchro2' := Synchro2 ||
    Synchro2'ToBeUpdated := FALSE ||
    SynchroEnabler := FALSE
END;
enableSynchro = SELECT
        SynchroEnabler = FALSE
    THEN
        SynchroEnabler := TRUE
END;
disableSynchro = SELECT
        SynchroEnabler = TRUE
    THEN
        SynchroEnabler := FALSE
END
END
```

5. BIBLIOGRAPHY

1. VSIA, : System Level Interface Behavioural Documentation Standard (v1.1.0)
2. Abrial J.R: Event Model Decomposition (2002)
3. Abrial J.R: Guidelines to Formal System Studies (2000)

Chapter 10

REFINEMENT OF FINITE STATE MACHINES WITH COMPLEMENTARY MODEL CHECKING

Alexander Krupp, Wolfgang Mueller
Paderborn University/C-LAB, Paderborn, Germany

Abstract: We investigate a methodology for the refinement of finite state machines to time annotated finite state machines. We present a translation, which transforms models into B language code for efficient application of theorem proving. Refinement and verification is performed across three levels with the B-Method and with complementary verification through the RAVEN model checker. The presented methodology is outlined by the example of an echo cancellation unit in Chapter 14.

Key words: Formal refinement, model checking, state diagram, finite state machine.

1. INTRODUCTION

HW/SW co-design typically starts with a functional model (e.g., a C program), which is transformed into a cycle-accurate finite state machine (FSM) oriented model, from which HW and SW components are derived. In later design steps, the FSM is annotated with timing information for the verification of its real-time behaviour. Here, annotation is typically not manually introduced rather than being inserted by backannotation from a first synthesis. Several RT-level synthesis tools, for instance, support backannotation of timing information through the SDF standard.

As a complementary approach to classical simulation, formal verification (i.e., equivalence checking, model checking, SAT solving) plays an increasingly important role in the domain of HW/SW co-verification, where most of the approaches are based on the verification of state oriented models. Symbolic model checking, for instance, verifies if a property specification

J. Mermet (ed.), UML-B Specification for Proven Embedded System Design, 155–167.
© 2004 *Kluwer Academic Publishers. Printed in the Netherlands.*

given by a temporal logic formula is true with respect to a model given by synchronously communicating finite state machines, which are most oftenly based on the theory of Kripke structures. For those models, state diagrams[1] are taken for documentation and for graphical capture.

In this chapter, we present a design subflow for proven design based on the refinement of finite state machines. The chapter describes an efficient combination of theorem proven refinement and model checking based on the RAVEN model checker [11] and the Atelier-B B toolset [1]. We have selected the RAVEN model checker, since RAVEN is based on an extension of Kripke-Structures for the verification of time annotated real-time models and timed property specifications, so that it can manage the verification of real-time systems without any modification. In our verification environment, we apply state diagrams as a graphical capture for the RAVEN Input Language (RIL). In a first step, we can validate the cycle accurate models by means of simulation and formally verify properties through model checking by RAVEN. After first verifications, we automatically generate corresponding B models (i.e., B machines) to prove the correctness of refinements undertaken in later design steps. Here, our focus is on the refinement of the synchronous RIL model through annotation of transitions with timing specification and removal of non-determinism.

Figure 1 gives an overview of the verification flow from state diagrams to B for formal refinement and final C code generation.

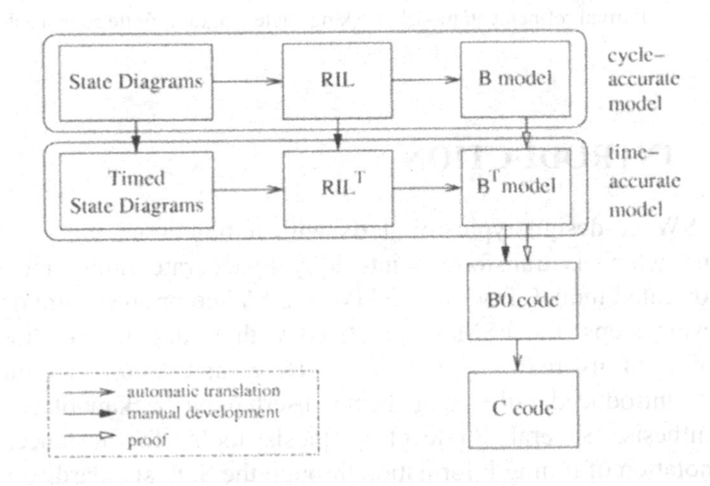

Figure 1: Combined Model Checking and Formal Refinement

In details, at cycle accurate modelling level, RIL code is generated from the state diagram for the RAVEN model checker. The RIL code is then

[1] including StateCharts, UML 1.5 state diagrams, and UML 2.0 state machine diagrams

transformed into an equivalent B code representation for formal refinement. After refining the state diagram by removal of non-determinism and insertion of timing information, with the same transformation, we arrive at corresponding RIL and B, which we denote as RIL^T and B^T, respectively. Now, the B prover can be applied to verify that the B^T model is a correct refinement of the B model, which consequently verifies that the RIL^T model is a correct refinement of the RIL model as well. After successful verification, B tools can be applied to generate a B implementation in B0, from which Atelier-B can easily generate Ada or C code.

The deviation trough RIL gives two main benefits. Firstly, we can apply simulation and model checking with deadlock analysis etc. for complementary verification without any extra code translation efforts. Secondly, through the direct mapping of RIL to B, we enforce a FSM oriented representation of the B code, which can be applied for efficient refinement automation. For the latter, we have developed B patterns, for which we have demonstrated that automatic theorem proving in combination with low runtimes is feasible when applying the Atelier-B theorem prover. Though we can presently demonstrate full automation with low runtimes just for two case studies, we think that we are heading in the right direction for the general adoption of our methodology. This would be a real advance for refinement automation, since the lack of provision of fully automatic proofs is still the main obstacle for the wide industrial acceptance of theorem proving; current systems still require too many user interactions.

In the next section, we briefly present what is currently available in the field of approaches combining theorem proving and model checking. Thereafter, we introduce RAVEN and its input language RIL. The main section outlines basic concepts of B refinement and introduces a three level B refinement scheme which our verification flow is based on.

2. COMBINED MODEL CHECKING AND THEOREM PROVING

Several systems are available, which integrate model checkers into theorem provers or vice versa.

PVS (Prototype Verification System) is a theorem prover where the PVS specification language is based on high order predicate logic. Shankar et al. enhance PVS with tools for abstraction, invariant generation, program analysis (such as slicing), theorem proving, and model checking to separate concerns as well as to cover concurrent systems properties [12]. STeP (Stanford Temporal Prover) is implemented in Standard ML and C [9] and integrates a model checker into an automatic deductive theorem prover. The input for model checking is given as a set of temporal formulae and a transition system, which is generated from a description in a reactive system specification language (SPL) or a description of a VHDL subset. The SyMP framework integrates a model checker into a HOL based theorem prover for general investigations on effectiveness and efficiency [4]. There, focus is on computer assisted manual proofs where main examples come from hardware design. Mocha [2] is a model checker enhanced by a theorem prover and a simulator to provide an interactive environment for concurrent system specification and verification. However, in the Mocha framework, theorem proving is interactive and no efficient reasoning of more complex systems is reported. In the context of B, Mikhailov and Butler combine theorem proving and constraint solving [10]. They focus on the B theorem prover and the Alloy Constraint Analyser for general property verification. Fokkink et al. employ the B method and combine it with μCRL [3]. They describe the use of B refinement in combination with model checking to arrive at a formally verified prototype implementation of a data acquisition system of the Lynx military helicopters. They present the refinement of a system implementation starting from a first abstract property specification. All those approaches consider timeless models and do not cover refinement with respect to real-time properties in finite state machines. Only Zandin has investigated real-time property specification with B by the example of a cruise controller [13]. However, he reports significant problems with respect to the complexity of the proof during refinement.

In contrast to the HOL based approaches, we present a model checking based approach with the RAVEN model checker in combination with the Atelier-B theorem prover for formal refinement with very little user interaction. We focus on the verification of finite state machine based real-time systems and their refinement to time accurate models based on an efficient mapping from the RAVEN Input Language to the B language.

3. RAVEN

A model checker verifies a given set of synchronously communicating state machines with respect to properties given by a set of formulae in tree temporal logic, mostly in CTL (Computational Tree Logic) or in LTL (Linear Tree Logic) [6].

For our approach, we apply the RAVEN (Real-Time Analyzing and Verification Environment) real-time model checker, which extends basic model checking for real-time systems verification by additional analysis algorithms [11]. In RAVEN, a model is given by a time-annotated state transition system, i.e., a set of I/O-Interval structures [11] where the property specification is given by a CCTL formula. I/O-Interval structures are based on Kripke structures with [min,max]-time intervals at their state transitions. We only briefly sketch the basics of interval structures here before we introduce the RAVEN Input Language (RIL) by some examples.

An interval structure $\mathfrak{I}$ is a tuple (P, S, T, L, I) with a set of propositions P, a set of states S, a transition relation T between states such that every state has a successor state, a state labelling function $L: S \rightarrow wp(P)$, and a transition labelling function $I: T \rightarrow wp(N)$. I/O-interval structures are interval structures that have read-only access to states of other interval structures. We assume that each interval structure has exactly one clock for measuring time. The clock is reset to zero, if a new state is entered. A state may be left, if the actual clock value corresponds to a delay time labelled at an outgoing transition. The state must be left when the maximal delay time of all outgoing transitions is reached.

Clocked CTL (CCTL) is a time-bounded temporal logic [11]. In contrast to classical CTL, the temporal operators **F** (i.e., eventually), **G** (globally), and **U** (until) are provided with interval time-bounds $[a,b]$, $a \in N_0$, $b \in N_0 \cup \infty$, where ∞ is defined by: $\forall i \in N_0 : i < \infty$. The temporal operators may have a single time-bound only. In that case, the lower bound is set to zero by default. The lower bound is zero and the upper bound is infinity by default. The **X**-operator (i.e., next) may have a single time-bound a ($a \in N$) only. If no time bound is specified, it is implicitly set to one.

The semantics of CCTL is defined as a validation relation $\models$, using the notion of runs, which represent possible sequences of clocked states that occur during execution of $\mathfrak{I}$. Any arbitrary clocked state g_0 may be the starting point of a run. Table 1 shows some sample semi-formal descriptions of the validation relation for a given interval structure $\mathfrak{I}$ and a clocked state $g_0 = (s_0, v_0) \in G$. Note that ϕ and ψ both stand for arbitrary CCTL formulae. The semantics for temporal operators with path quantifier **A** (i.e., regarding all possible runs) can easily be derived. E.g., $\mathbf{AX}_{[a]}\ \phi$ is equivalent to $\neg\mathbf{EX}_{[a]}\ \neg\phi$ and $\mathbf{AF}_{[a,b]}\ \phi$ is equivalent to $\neg\mathbf{EG}_{[a,b]}\ \phi$.

In the context of RAVEN, I/O-Interval structures and a set of CCTL formulae are specified by means of the textual RAVEN Input Language (RIL). A RIL specification contains

(a) a set of global definitions, e.g., fixed time bounds or frequently used formulae,

(b) the specification of parallel running modules, i.e., a textual specification of I/O-Interval structures, and

(c) a set of CCTL formulae, representing required properties of the model.

Formula	Denotation	Description
$g_0 \models p\ (p \in P)$	Proposition	g_0 is valid in p, if $p \in L(s_0)$
$g_0 \models \neg\phi$	Negation	g_0 is satisfied by $\neg\phi$ if $g0 \models \neg\phi$ is false.
$g_0 \models (\phi \wedge \psi)$	Concatenation	$g_0 \models \phi$ and $g_0 \models \psi$ \
$g_0 \models EX_{[a]}\ \phi$	Next	There exists a run $r = (g_0, \ldots)$ such tha$_t$ $g_0 \models \neg\phi$
$g_0 \models EF_{[a,b]}\ \phi$	Eventually	There exists a run $r = (g_0, \ldots)$ and $a \leq i \leq b$ such that $g_i \models \phi$
$g_0 \models EG_{[a,b]}\ \phi$	Globally	There exists a run $r = (g_0, \ldots)$ and $a \leq i \leq b$ such that for all $g_i \models \phi$
$g_0 \models E(\phi\ \underline{U}_{[a,b]}\ \psi)$	Strong Until	There exists a run $r = (g_0, \ldots)$ and an $a \leq i \leq b$ such that $g_i \models \psi$ and for all $j < i$ holds $g_i \models \phi$
$g_0 \models E(\phi\ U_{[a,b]}\ \psi)$	Weak Until	There exists a run $r = (g_0, \ldots)$ and either (a) there exists an $a \leq i \leq b$ such that $g_i \models \psi$ and for all $j < i$ holds $g_i \models \phi$, or (b) and for all $j \leq b$ holds $g_i \models \phi$

Table 1: CCTL Overview

If RAVEN evaluates a CCTL formula to be incorrect, a counter example execution run can be generated. Execution runs are given by time-annotated sequences of state changes, which can be displayed by a built-in waveform browser that lists all variables and their states over time.

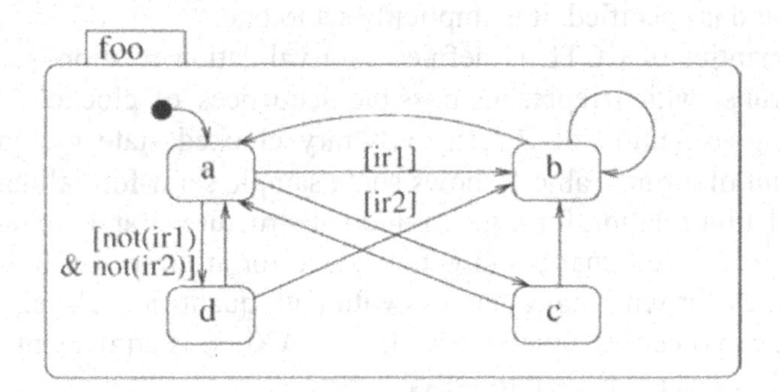

Figure 2: State Diagram for `foo`

To give an overview of RIL, we consider the small example in Figure 2. Figure 3 gives the corresponding RIL code, which defines one I/O-interval structure within one module.

The example has four states `a`, `b`, `c`, and `d`. For state `a`, we have defined three transitions. The first two are triggered by the values of `ir1` and `ir2`. The third one is the default transition for ¬`ir1` and ¬`ir2`. From states `b`, `c`, or `d` transitions fire non-deterministically to either state `a` and `b`. Note that, once in state `b`, due to one self-loop, the execution may remain in that state.

```
MODULE foo
TRANS
       |- s=a -- ir1 --> s:=b; r:=r - 1
              -- ir2 --> s:=c; r:=r + 1
                     !-> s:=d
       |- s=b --     --> s:=a
              --     --> s:=b
       |- s=c --     --> s:=a
              --     --> s:=b
       |- s=d --     --> s:=a
              --     --> s:=b
END
```

Figure 3: RIL Code for foo

Module foo may be refined to the RIL^T module bar by replacing the non-deterministic transition from state b to a and b

```
|- s=b --      --> s:=a
               --> s:=b
```

by a timed transition with delay 5:

```
|- s=b -- :5 --> s:=a
```

This means that after 5 time steps after s=b, the value of s changes from b to a. The corresponding state diagram notation is shown in Figure 4

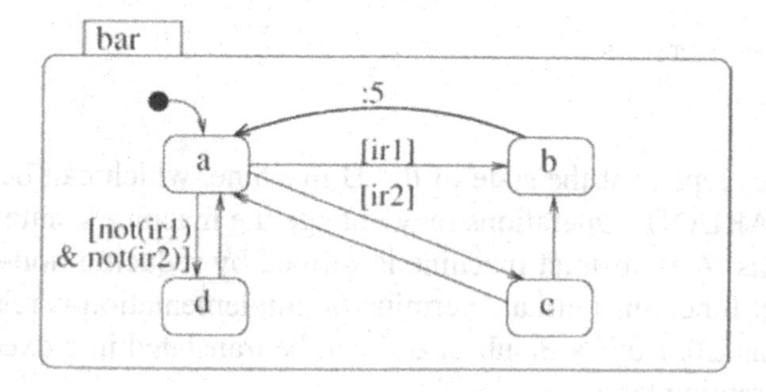

Figure 4: State Diagram for bar

4. REFINEMENT OF FSM-BASED REAL-TIME SYSTEMS

This section first introduces the basic principles of B refinement before outlining our approach to the refinement of finite state machines.

4.1 Refinement in B

The two classical approaches to theorem proving in the domain of electronic design automation are the Boyer--Moore Theorem Prover (BMTP) and HOL [5,8]. BMTP and HOL are both interactive proof assistants for higher order logic. In theorem proving, a proof has to be interactively derived from a set of axioms and inference rules. Though several practical studies have been undertaken, classical interactive theorem proving has not received a wide industrial acceptance. B was introduced by Abrial in [1] as a methodology, a language, and a toolset.

Similar to Z, B is based on viewing a program as a mathematical object and the concepts of pre-and postconditions, of non-determinism, and weakest precondition. The B language is based on the concept of abstract machines. An abstract B machine consists of VARIABLES and OPERATIONS and is defined as follows

```
MACHINE M( ... )
 CONSTRAINTS
 ...
 VARIABLES
 ...
 INVARIANT
 ...
 INITIALISATION
 ...
 OPERATIONS
 ...
```

Variables represent the state of the B machine, which can be constrained by an INVARIANT. Operations may change the machine's state and return a list of results. A B abstract machine is refined by reducing non-determinism and abstract functions until a deterministic implementation is reached, which is denoted as B0. B0 is a B subset and can be translated into executable code of a programming language.

Refinement in B means the replacement of a machine M by a machine N, where the operations of M are (re)defined by N. Syntactically, a refinement is given by

```
REFINEMENT N REFINES M
...
```

In N, M operations have to be given by identical signatures. However, they may refer to different internal states or even to a different definition. The important requirement is that the user must be able to use machine N as she would use machine M. Refinement can be conducted by

(i) removal of preconditions and non-deterministic choices,
(ii) insertion of control structures, e.g., sequencing and loops, and
(iii) transformation of data structures.

A final refinement to B0 defines the implementation of the system as

```
IMPLEMENTATION I REFINES M
...
```

At implementation level, the B0 definition looks very much like a PASCAL program, from which a B toolset can automatically generate Ada, C, or C++.

4.2 FSM Refinement

Based on the principles of B, we elaborate on refinement from structural to behavioural cycle-accurate finite state machines, and furtheron to time-accurate state machines. For this, we distinguish three levels of B refinement: Structural (S), Behavioural (SB), and Timed Behavioural (SBT) in order to structure the B machines according to their refinement level and to conduct efficient proofs with the Atelier-B theorem prover (cf. Figure 5).

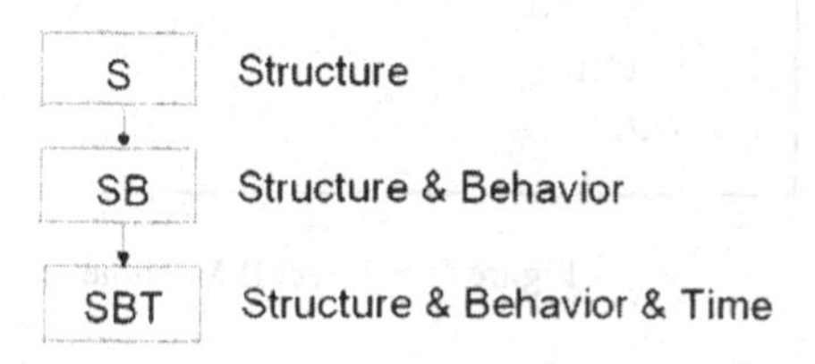

Figure 5: Different Levels of Refinement for Real-Time Systems

To distinguish B the corresponding machines, we apply suffixes S, SB, and SBT to the B machine identifiers.

The first B refinement level denoted as S-level is introduced to capture structural properties and generic behaviour through non-deterministic state transitions. SB-level machines cover non-deterministic behaviour, where the SBT-level refers to deterministic time-accurate definitions. The following outlines are based on the foo RIL module from the previous section.

4.2.1 Structural Level

For the structural (S-level) specification, we apply B specifications with:

- signal definitions and their relations given as invariants,
- an operation for the propagation of values between signals, and
- an operation, which implements the model of computation (MoC) to trigger state transitions of given the FSMs: doTransition

The below B code sketches a B machine at S-level (foo_S) operating for a state SIGNAL_s of a type defined by set SIGNAL_T_s. The operation doTransition implements a transition to any possible state from SIGNAL_s, by a non-deterministic choice over all possible values of SIGNAL_s. Thus, any transition under the given invariant can be selected. The ANY..WHERE..END clause is basically equivalent to the predicate $\forall ss: ss \in SIGNAL\_T\_s \Rightarrow [SIGNAL\_s := ss]$.

```
MACHINE foo_S
...
INVARIANT
   SIGNAL_s : SIGNAL_T_s ...
OPERATIONS
 ...
  doTransition =
  BEGIN
    ANY ss ...
    WHERE ss:SIGNAL_T_s ...
    THEN SIGNAL_s:=ss
    END
  END
END
```

Figure 6: S-Level B Machine

The above definition gives a basic structure, whose implementation is given at SB-level, which is introduced hereafter.

4.2.2 Behavioral Level

The behavioural level (SB-level) B machine adds behavioural properties to the previous structural definition by mainly refining the operation `doTransition`. Refinement in B is conducted by:

- removal of non executable elements of the pseudo-code (pre-condition and non-deterministic choice),
- insertion of transition details and classical control structures of programming (sequencing and loop), and
- transformation of data structures (sets, relations, functions,...).

In our previous S-level specification, a transition from every state to any other state was specified as a non-deterministic transition. In the refined part of the specification, transitions are restricted - to some extent - to deterministic transitions. This is always a refinement from S-level, because we restrict the postcondition of the transition operation, by removing some of the S-level transitions. We have separated the B machines into S-level and SB-level to significantly reduce the number of (tool-generated) proof obligations for proof automation.

```
REFINEMENT foo_SB
REFINES foo_S
...
  doTransition =
  BEGIN
    IF SIGNAL_s = aa THEN
      SELECT (DEFINE_ir1=TRUE) THEN
        ANY ... WHERE ss=bb ... THEN ... END
      WHEN (DEFINE_ir2=TRUE) THEN
        ANY ... WHERE ss=cc ... THEN ... END
      ELSE ANY ... WHERE ss=dd ... THEN ... END
      END
    ELSIF SIGNAL_s = bb THEN
      SELECT (TRUE=TRUE) THEN
        ANY ... WHERE ss=bb ... THEN ... END
      WHEN (TRUE=TRUE) THEN
        ANY ... WHERE ss=aa ... THEN ... END
      END
   ...
```

Figure 7: SB-Level B Machine

The previous B code shows the B representation of RIL module foo of Figure 3. The transition relation is represented in a decision structure with 3 nested statements. The outer `IF..ELSEIF..END` structure matches the current machine state with the source state of a transition. Therein, the SELECT..WHEN..ELSE..END selects substitutions depending on the guard value of the Boolean transition. A default transition may be specified in the ELSE branch. Finally, ANY statements select the states. Note here, that the SELECT (TRUE=TRUE) statement is a straightforward translation of the RIL empty input restriction, which in RIL equals to TRUE:1.

4.2.3 Timed Behavioral Level

At timed behavioural level (SBT-level), the B machines capture structural, behavioural, and timing properties. This is achieved through refinement of the SB-level operation doTransition. Non-determinism is replaced by either providing a deterministic choice for a transition or by introducing a timed transition. Timed transitions are specified with an auxiliary variable encapsulated in a Timer B machine. For each timed transition, one Timer machine has to be included. For timed transitions an additional construct is inserted for the management of the timer. An IF...THEN...ELSE...END statement controls if the timer is advanced or reset. If the timer elapses, an IF..END statement executes a transition in a simultaneous substitution.

```
REFINEMENT foo_SBT
REFINES foo_SB
INCLUDES t1.Timer(5)
...
  doTransition =
  BEGIN
    ...
    ELSIF SIGNAL_s = bb THEN
      IF (t1.elapsed = TRUE)
      THEN t1.doReset
      ELSE t1.doAdvance END                           ||
      IF (t1.elapsed = TRUE) THEN
        ANY ... WHERE ss=aa ... THEN ... END
      END
    END
    ELSIF SIGNAL_s = cc THEN
...
```

Figure 8: SBT-Level B Machine

The previous B code shows the SBT-level refinement of foo_SB. In that code, the transition for state SIGNAL_s = bb is refined to a timed transition with delay 5. For this, two simultaneous substitutions are introduced to the corresponding `SELECT` statement. The first one updates the timer and the second one performs the state transition as soon as the timer elapses. The refinement from SB- to SBT-level described here compares to the refinement described for RIL modules `foo` and `bar` in the previous section. Note, that we have not used time constants rather than the explicit integer, i.e., `5`, to keep the previous code simple.

ACKNOWLEDGEMENTS

We appreciate the fruitful cooperation with the partners of the PUSSEE project and the help and support of Jürgen Ruf, Thomas Kropf, and Stephan Flake with the RAVEN model checker.

REFERENCES

1. J.R. Abrial. The B-Book. Cambridge University Press, 1996.
2. R. Alur and T.A. Henzinger. Reactive Modules. In LICS'96, 1996.
3. S. Blom et al. μCRL: A Toolset for Analysing Algebraic Specifications. In Proc. of CAV'01, 2001.
4. S. Berezin. Model Checking and Theorem Proving: A Unified Framework. PhD thesis, Carnegie Mellon University, 2002.
5. R.S. Boyer and J.S. Moore. A Computational Logic Handbook. Number 23 in Perspectives in Computing. Academic Press, 1988.
6. E.M. Clarke and E.A. Emerson. Design and Synthesis of Synchronization Skeletons Using Branching Time Temporal Logic. Lecture Notes in Computer Science, 131, 1981.
7. E.M. Clarke and W. Heinle. Modular Translation of State diagrams to SMV. Technical Report, Carnegie Mellon University, 2000.
8. M.J. Gordon. Introduction to HOL. Cambridge University Press, Cambridge, 1993.
9. Z. Manna et al. STeP: The Stanford Temporal Prover. Technical Report, Stanford University, 1994.
10. L. Mikhailov and M. Butler. An Approach to Combining B and Alloy. In D. Bert et al., editors, ZB'2002, Grenoble, France, 2002.
11. J. Ruf. RAVEN: Real-Time Analyzing and Verification Environment. J.UCS, Springer, Heidelberg, 2001.
12. N. Shankar. Combining Theorem Proving and Model Checking Through Symbolic Analysis. In CONCUR 2000, 2000.
13. J. Zandin. Non-Operational, Temporal Specification Using the B Method – A Cruise Controller Case Study. Master's Thesis, Chalmers University of Technology and Gothenburg University, 1999.

Chapter 11

ADAPTIVE CRUISE CONTROL CASE STUDY DESIGN EXPERIMENT

Fredrik Bernin[1], Michael Lundell[1], Ola Lundkvist[1] and Denis Sabatier[2]
[1]*Volvo Technology Corporation;* [2]*ClearSy*

1. INTRODUCTION

This chapter documents experience from a case study of adaptive cruise control (ACC). The chapter starts with a brief description of the model developed in the case study and an overview of the work performed in the case study. The main part of the chapter is organised in sections covering different aspects and issues encountered in the case study.

2. MODEL OVERVIEW

The model in this case study is made at two levels, a specification of user level requirements and a refinement that decomposes the system to sub-systems.

2.1 User level model

The user level model is designed from a user's point of view, which means that it models the behaviour of an ACC as the user experience it. Each implementation of an ACC can be evaluated by comparing the implementation to the user level model and if the evaluated ACC has the properties that are specified in the model, it will be considered to be correct.

By only modelling the behaviour of an ACC, no restrictions are made to the implementation of the ACC. For example, it might be radar, laser, or

J. Mermet (ed.), UML-B Specification for Proven Embedded System Design, 169–197.
© 2004 *Kluwer Academic Publishers. Printed in the Netherlands.*

infrared detection of obstacles. The model states only that an obstacle of a certain size must be detected at a certain range and that the vehicle must change its behaviour according to the new environment within a certain time.

The specification consists of two parts: one that contains the model written in B and another part with a dictionary that explains all the events and variables in the B model. The dictionary is necessary to give meaning to the B model.

The model consists of a number of events that modifies the different variables. The model only uses environmental or physical variables such as `ObstacleDistance` (the distance, usually referred to as a time interval, to the current obstacle), and `VehicleTryKeepSpeed` (boolean which is true when the vehicle really is trying to reach and keep a certain cruise speed). All these variables can be observed in the real world. All variables in the B model are boolean or set values, stating conditions of physical values. The actual obstacle distance is for example not modelled, only the fact that the obstacle is close or far away is modelled.

It is obvious that an observer can examine `ObstacleDistance`, but it might be harder to examine `VehicleTryKeepSpeed` objectively. To be able to judge whether the vehicle actually tries to reach and keep the cruise speed, `VehicleTryKeepSpeed` is defined mathematically by using other variables that can be objectively examined. An example of such definition could be: if the acceleration of the vehicle is between $1m/s^2$ and $3m/s^2$ when the `VehicleSpeed` is less than the `CruiseSpeed`, `VehicleTryKeepSpeed` is true. These formulas are defined outside of the formal model.

There exist two kinds of events. The first type is environment events like `SetCruiseSpeed` (the cruise speed button is pressed) and `ObstacleDisappears` (an obstacle disappears). These events are the most common and describe environmental changes that may occur.

The second type can be called Timing events and is used to deal with timing requirements. There are two timing requirements in the user level model and each requirement has a corresponding event. When the environment is changed in such way that the vehicle must change behaviour in a specified time, the environment event sets a flag. That flag has a corresponding timer event that will occur after a specified time after the flag been set. If the behaviour of the vehicle has not changed when the timing event occurs, it will be a conflict because the guard of the timing event contains all relevant parts of the invariant. These events does of course not occur in reality, but it is easy t imagine them as the events of starting, checking and stopping a timer to verify the timing requirements.

The user level model is quite simple and easy to understand. The model contains 10 variables and 17 events. The invariant contains the main ACC laws, which is to be preserved by the model. Figure 1 below shows the law, which should be easy to understand even though all variables are not explained here.

```
/* ACC user level laws */
(CruiseAllowed = FALSE => CruiseActive = FALSE) &
(CruiseActive = TRUE &
        ClosestObstacleChanged  = FALSE &
        CruiseActivationInProgress = FALSE =>
        /* do something law */
        (VehicleTryKeepSpeed = TRUE or
                VehicleTryKeepTimeGap = TRUE) &
        /* keep speed if far or no obstacle */
        (ObstacleDistance = ODfar =>
                VehicleTryKeepSpeed = TRUE) &
        /* keep time gap otherwise, if not too fast */
        (ObstacleDistance = ODclose &
                TimeGapAttitudeWouldBeMore = FALSE =>
                VehicleTryKeepTimeGap = TRUE
        ) &
        /* keep speed, if time gap keeping would be too fast */
        (ObstacleDistance = ODclose &
                TimeGapAttitudeWouldBeMore = TRUE =>
                VehicleTryKeepSpeed = TRUE
        )
)
```

Figure -1. User level model invariant

The user level model was made entirely in the B environment from the beginning, but at the end of the case study, the model was translated to a UML model. Figure 2 shows the class diagram of the user level model. Only two classes are used at this time, one for the Cruise and one for the closest obstacle present.

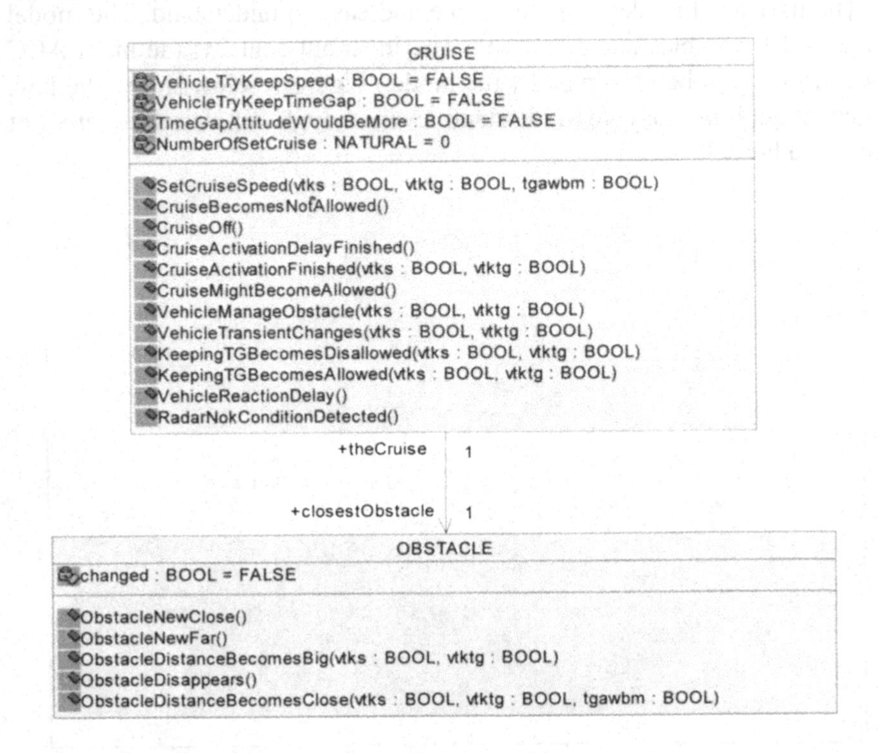

Figure -2. UML class diagram

Each of these classes has an associated StateChart. Figuree 3 shows the StateChart for the Cruise class and figure 4 shows the StateChart for the Obstacle class. Using the StateCharts, some variables are encoded in the states, and preconditions related to the states are defined by the possible transitions in the StateChart. Other preconditions are defined on the transitions or in the specification of each event.

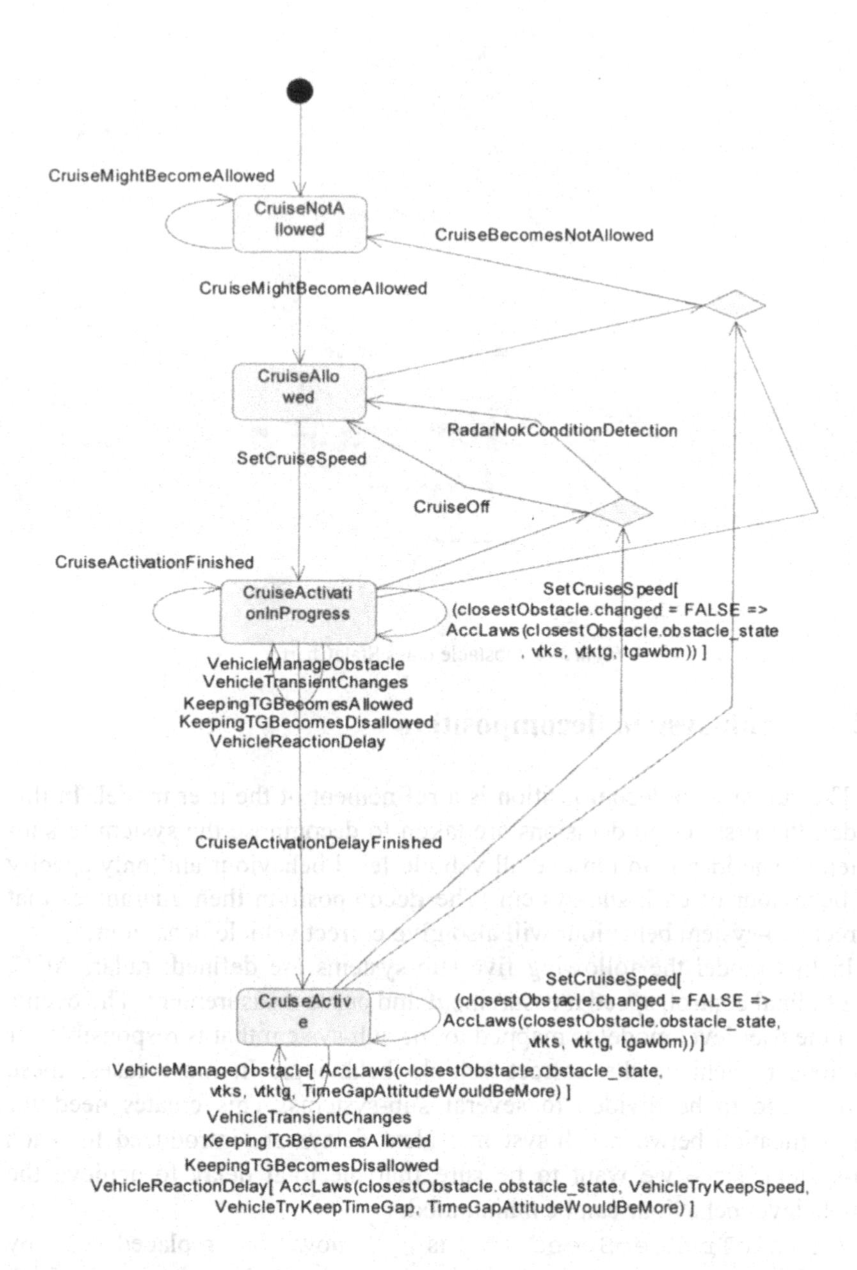

Figure -3. Cruise class StateChart

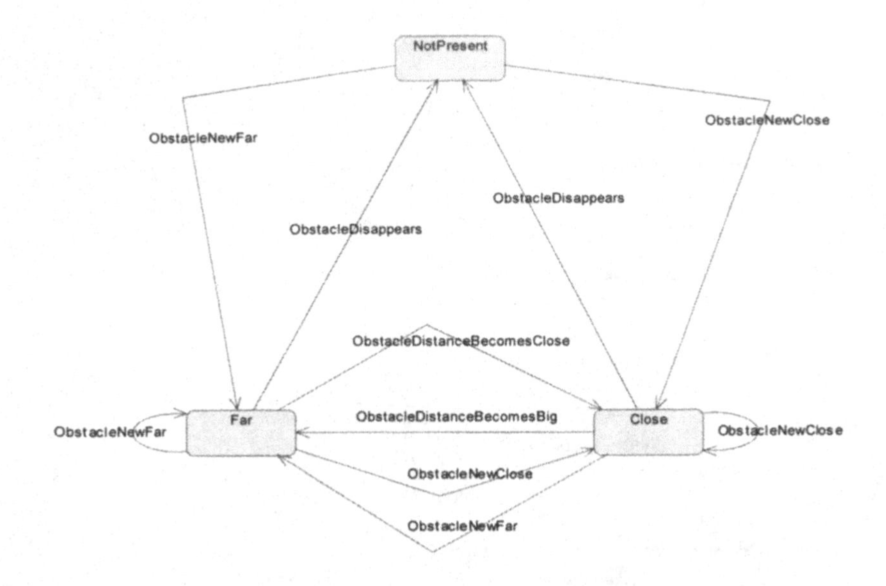

Figure -4. Obstacle class StateChart

2.2 Sub-system decomposition

The sub-system decomposition is a refinement of the user model. In this model, the first design decisions are taken to decompose the system to sub-systems. The idea is to remove all vehicle level behaviour and only specify the behaviour of each sub-system. The decomposition then guarantees that correct sub-system behaviour will also give correct vehicle behaviour.

In this model the following five sub-systems are defined: radar, ACC, longitudinal control, speed measurement and curve measurement. The events from the user level model is mapped to the sub-system that is responsible for reactions to achieve the correct vehicle behaviour. In most cases, these events need to be divided to several sub-systems. This creates need for communication between sub-systems. New delays are introduced for each sub-system, since we want to be sure that the total delay to achieve the vehicle level behaviour can be maintained.

`VehicleTryKeepSpeed` is now replaced by `ACCCalculatesTorque`, which represents the requested torque, which must correspond to an appropriate torque interval needed to reach and keep the cruise speed when `VehicleTryKeepSpeed is true.` Another introduction in this model is different faults and conditions for each sub-system that disallows the ACC from operating, since achieving the specified behaviour would not be possible.

The decomposition has not been fully completed. Some events need to be further split since they represent actions of more than one sub-system.

The decomposition is more complex than the user level model. The decomposition contains 22 variables and 24 events, and further variables and event are needed to complete the decomposition. The invariant in particular is quite more complex with new laws for each sub-system and gluing between replaced variables. Figure 5 presents a few examples from this invariant.

```
/* radar fast enough law */
(RadarReactionDelay = TRUE =>
    ClosestObstacleChanged = TRUE) &
(RadarTransient = TRUE => RadarReactionDelay = TRUE) &
/* only if not transient, object measures represent reality */
(RadarInfoOK = TRUE & RadarTransient = FALSE &
    ACCTransient = FALSE =>
    ACCEstimationObstacleDistance = ObstacleDistance &
    ACCTorqueTGWouldBeMore = TimeGapAttitudeWouldBeMore
) &
/* Gluing: as speed measures are always good,
    speed torque gives speed attitude */
(ACCCalculatesTorque = ACTspeed =>
    VehicleTryKeepSpeed = TRUE) &
/* Gluing: only if good object measures,
    distance torque gives time gap attitude */
(RadarInfoOK = TRUE & RadarTransient = FALSE &
    ACCTransient = FALSE &
    ACCCalculatesTorque = ACTdistance =>
    VehicleTryKeepTimeGap = TRUE
)
```

Figure -5. Decomposition invariant examples

A translation to a UML model has not yet been performed for the decomposition. It would be an interesting study though, since this model is more complicated and should better be able to utilise the use of several classes.

3. WORK OVERVIEW

The ACC models have been developed gradually in a few major steps:
1. Ordinary simple cruise controller
2. Simple ACC
3. Simple ACC with timing requirements
4. Decomposition to sub-systems by refinement

The first model was simple ordinary cruise controller, by adding events and variables an ACC without any timing requirements evolved. The timing events were added in the last step. Some simplifications still remain, since it would be too time consuming to try to add all requirements in this case study.

It was decided early that only Boolean variables and finite sets would be used to ease the burden of proof. The disadvantage of only using Boolean variables is that reality is not easily described in terms of true and false. One example is the vehicle speed that would naturally be described using an integer or a float, but in the model, several Boolean variables are used to represent the vehicle speed. This choice was made to keep the B models away from numerical models that are better dealt with specialized tools like MatLab.

Events were added by considering different use cases; each use case resulted in some new events that related to the use case. An example of a use case is following an accelerating obstacle. At start, the vehicle is keeping a distance to the obstacle, which is assumed to be close and moving slower than the wanted cruise speed. The vehicle accelerates, keeping the distance until the obstacle exceeds the wanted cruise speed. At this time, the vehicle changes its behaviour to keep the cruise speed instead.

The final user level model is quite simple to understand, but during the project the model was more complicated and more difficult to understand. Analysing new use cases and adding variables and events to the model was quite difficult, since effects on existing variables and events must always be considered. Reaching the final model was harder than what can be expected when looking at the model.

Decomposing the model required some attempts until a successful approach was found. Particularly, the user level model is quite different from typical software specifications, since we define events of the environment, which cannot directly be mapped to sub-systems. We also have a system with one major task, where most events cannot easily be allocated to different sub-systems. Most events need to be split to several sub-events, since many sub-systems cooperate to achieve the task of the system.

The decomposition has not been fully completed, but we are confident that the intermediated result is in the right direction. The decomposition model is more complex than the user level model, but when the decomposition is completed, the model could be split into five models, one for each sub-system. The model for each sub-system should then be at the same level of complexity as the user level model.

Proving was mainly performed when each model was finished. Since we did some iterations of each model, there were several proof sessions. For the user level model, these proofs were quite simple, which is not a surprise with such a simple invariant. The main work in these proof sessions was rather to correct and complete the model such that it could be proved. Most proofs were discharged automatically and only a few easy manual demonstrations were needed. The user level model generated 725 obvious proof obligations, 73 proof obligations that were automatically proved, and 2 proof obligations that were proved by manual demonstration.

Proving the decomposition is more complicated, which can be expected, given the invariant and particular the gluing links to the user level model. The decomposition was actually performed in two steps to simplify proofs. In the first refinement the variables to be replaced were kept, so that these could be utilised in the proofs. In the second refinement, these variables were finally dropped. The first refinement generated 2970 obvious proof obligations, 149 proof obligations that were automatically proved and 45 proof obligations that were proved by manual demonstration. The second refinement generated 3525 obvious proof obligations, 119 proof obligations that were automatically proved, 9 proof obligations that were proved by manual demonstration, and 1 proof obligation that has not yet been proved. As the figures tell us, many of the proof obligations for the refinements were quite tricky. In many cases, the problem is not that the proofs are particularly difficult, but that they are big because they involve quite a lot of variables. It could be interesting to study ways to improve the proof time and number of automatically proven proof obligations, for example by use of an efficient model checker.

The final activity in the case study was the remodelling to UML. Eventually this showed to be quite easy, and the model itself is much easier to understand. We think that we would have benefited from using the UML approach from the beginning, resulting in quicker progress of the case study, if the U2B tool would have been ready then.

In conjunction with the development of the UML model we also tried using a model checker and animator for B, which University of Southampton is developing in another project. This was a good way to examine the model to determine if it really represents what we want. We discovered a few mistakes and updated the model. We have not actually proved that the B generated from the UML model is consistent with the original B model, we have only tested a few scenarios in the animator. Considering that we have made corrections to the original B model when developing the UML model, we can probably assume that the UML model is at least as correct as the B model.

One concern with the UML model was that the initial generated B was much more complicated to prove due to the instances supported in object orientation. Instead of 75 proof obligations with only 2 that required manual proofs, we got 750 proof obligations with 150 that had to be proved manually. We investigated a little how this could be simplified by removing the use of instances, since we do not need that in this case anyway. This significantly reduced the number of generated proof obligations to about 200, which all proved automatically. However, this is not a good solution, since it prevents us from using many of the object-oriented ideas in UML. In general, use of instances is needed in a model, so this must be supported as the main approach. It is important to be able to utilise instances without a big penalty on the proof effort. Our case study shows that it may be necessary to implement some mechanism to improve capabilities of proving models that make use of instances.

We used the original requirements as an input in the beginning of the project, but we have not continuously follow up them during our work. Afterwards we have analysed the inclusion of these requirements and updated our specification. Some requirements are still skipped due to simplifications we have made to reduce the amount of work. Now, when the approach is known, we would emphasize definition of scenarios early to cover all requirements. In this way, we can be more certain to really cover all requirements. In the case study, the use of scenarios has been more stressed at a later stage, when we discovered that all requirements were not sufficiently covered. The scenarios in the case study are in textual form, but they should be easy to arrange as use cases in UML.

4. DESIGN EXPERIENCE

This chapter contains different issues and problems that occurred or were identified during the project. Each issue is described, consequences and constraints are discussed, solutions and workarounds are proposed if any such has been identified.

4.1 Methodology and process

During this project, there has been little or no reasoning about what problem that we need to solve and how we should approach these problems. There is in fact no focus on methodology. The project has been quite diversified with focus on tool support for different translation and transformation methods, but the ideas of how to use these techniques to solve real problems has been quite vague and has not been sufficiently addressed. In the case study we had initially little concern on process and methodology, assuming that these would easily emerge in the project. This was later a concern, but at the end our main focus is modelling techniques and, for good and worse, we more or less leave process and methodology out of the scope for this project.

The aim was initially to go from system specification to implementation, though only a small part of the system would reach the implementation level. This would cover a complete development process, and we would achieve what seems to be the core of the PUSSEE/B method, a proven implementation. After some time we realised that we would not manage to reach implementation level. As a result, we have to think about the purpose of using the B method, and to fit this work in a complete process, because in our case it is perhaps not to reach a proven implementation that is the purpose. The user level model and the decomposition model have a value on their own, but how do we proceed then? Is there a path in the PUSSEE methodology that we could follow or do we go back to a traditional process? Our impression is that for this kind of system, using the PUSSEE methodology to implementation level is too difficult and would require too much effort. The probable approach would be to choose not to use formal refinements all the way to implementation.

Lack of methodology and process has made the work in the case study more difficult. It is essential to have a clear view of the purpose with each model, to be able to efficiently make useful models. It is also essential to have some process, with defined activities, that get you on the right track and guide you towards the goal. This case study has been an experiment not only involving specification languages, but also the overall work approach, so it is natural to experience these kinds of problems. However, more systematic work on a process from the beginning would probably have been beneficial to the case study.

4.2 Modeling methods

4.2.1 Abstract specification

The modelling approach used in the case study has mainly been communicated on work meetings. The approach is based on identifying events and variables by analysing scenarios. The events are very similar to use cases; they represent what can happen to the system. The variables are our model of the system. Each event is analysed to define which variables that are affected by the event. Each variable is also analysed to identify which events that can modify the variable, possibly identifying new events.

This approach is quite easy to understand, but still it created a lot of confusion in the beginning. Particularly we were missing some guidelines on modelling methods. We relied too much on guidance and demonstrations from B experts. This made it difficult for us to get started with the case study in the beginning.

A difficulty when making the model is to decide which variables and events to include in the model. Particularly we should note that making abstract models is in general much more difficult than a typical design model, which is more concrete. We can expect that making an abstract model for this type of system is rarely a straightforward process; it may be necessary to experiment with alternative ways of modelling. As in the case study, this may result in removing and remaking parts of the model, which may seem as a waste of work. We do not think that this can be entirely avoided, but maybe design patterns and user guidelines can be an aid to get on the right track as quickly as possible and reduce the amount of work that seems unnecessary.

The B method stresses the use of refinements, but we chose to make this specification at one level only, without refinements. Although we built the model gradually, the intermediate steps were not kept as refinements.

One motivation for this approach is that adding to the model often requires changes to what has been previously modelled. The effect from this is that you may need to change several steps of refinement, and to redo all proofs for these refinements. We avoided this problem, but instead our problem was that the specification soon got quite complicated. Using refinements would probably have helped us in getting a simpler model at the most abstract level. We do not know which choice is the better, but choosing the right levels of refinements could probably help in reducing these problems.

4.2.2 Refinement and decomposition

The refinement is a decomposition to five sub-systems. The first task of the decomposition is to identify appropriate sub-systems. Then events and variables are mapped to sub-systems. Many events and variables that define system behaviour need to be refined into smaller events and variables that are local to a sub-system. Each event in the user level model is analysed to identify how it affects each sub-system. These events are replaced by refined versions. In a similar way, variables that describe system properties are replaced by refined version that describe corresponding properties for each sub-system. Some events and variables represent things in the environment that do not decompose naturally with the sub-systems, but that we still can and need to observe. These are kept unmodified.

It was difficult to find a systematic approach to make this decomposition. It is not so clear from the beginning how to decompose events. You have to keep track on which sub-systems that are affected by an event, because it is of course still possible to define events that span over several sub-systems and that cannot be allocated to a single sub-system. The interfaces between sub-systems have so far not been particularly evident. One complication is also that many problems are not discovered until the model is being proved, which is generally performed when the model is finished, and it can then be necessary to remake the model.

Another difficulty is that refinement of events usually introduces intermediate states that may violate the invariant in the previous model. The invariant of a refinement usually becomes much more complicated for this reason. In some cases, it may be necessary to update the top level model to allow for a refinement. Care should be taken to avoid introducing exceptions to the invariant in the top level model and to define few and general events. Details and complexity should only occur in refinements, but also here you need to avoid particularly a too complicated linking invariant.

If you refine an event to a sequence of events, typically only the last event will be a refinement of the original event, since it is the only event that gives a result equal to the original event. Other events in the sequence are either new events, or refines some more general intermediate event. Sometimes this is a little annoying, because you would like to view the entire sequence as a refinement, which can be interleaved by other events, of the original event. This is in conflict with some very basic mechanisms of the B method, and such a change of semantics would most certainly not be possible in practice since proofs would be much more complicated.

4.3 Modeling a physical world

4.3.1 Representing formulas with discrete abstract variables

As we decided to make a user level model, we try to describe and reason about things in a physical world. In our case we are mainly concerned with vehicles and their properties such as speed and acceleration. When modelling this using the B method, we have several problems to consider.

We want to model continuous real variables such as speed, so we need to make an abstraction to be able to model them using the B method. A natural choice may seem to use discrete, non-continuous integer values. However, before making such a choice we should consider that a B model should be such that it must be possible to decide whether the implemented device really behaves according to the model. Such direct model interpretation requires link rules between abstract variables and events, and observed facts. If we would use integers, we would have difficulties defining this link rules between the abstract integer variables and the continuous physical values.

B asserts that if the implemented device behaves according to a B refinement, if this B refinement is proved against upper B levels in a way that preserves the link rules between abstract entities and observed facts, then the implemented device behaves according to the B upper level.

If we use integer abstract variables to represent continuous physical values, refinement will only prove that the composition of the integer calculus in the refinement corresponds to the integer calculus done in the above level. To evaluate the implemented device, one must check that the composition of the formulas for each part of the device gives the required formula for the whole device, and that given the accuracy of each part, the whole device has the required accuracy. This accuracy check cannot be done in a B refinement.

An additional problem when using integers in this way is that the proof effort is very high, due to algebraic proofs that must be made by hand, the tool's prover not being optimised for them.

It is instead necessary to use boolean abstract variables to represent the fact that a certain algebraic formula holds between concerned physical values. Such booleans are defined by giving the associated formula, with accuracy for each value. Through refinement, new booleans are defined for new concerned physical values (internal to the device). Low level booleans must be linked to upper level ones (gluing invariants). These gluing invariants will give the algebraic calculations to do to check that the final device is correct. All algebraic calculations are thus kept away from B, but the B development gives the composition rules to apply.

4.3.2 Choosing the right boolean or enumerated variables

Given any device which requirement is to keep a certain algebraic equation true between physical things, we could always specify its requested behaviour with the following:

```
MACHINE
  CorrectButUseless
VARIABLES
  Ok
INVARIANT
  Ok = TRUE
END
```

Where "Ok" is a boolean variable meant to be TRUE if and only if all the required physical equations characterizing the device as being correct are established. In "all equations", we include power supply, button presses, etc. This is a way of saying "do everything outside B".

If we want to make a profitable use of B, then we must use variables representing smaller parts of the physical requested laws. Thus the B modelisation and proof will account for how these small parts are linked, and how the specified device should comply to which laws depending on modes and important events.

The problem is: how to choose the set of B variables in a way that will lead to simple algebraic formulas for each variable? If the variable set is ill chosen, then several B variables representing the same complex formulas, with a useless splitting.

If an expert has to describe the requested behaviour of the device using algebraic formulas, he will write a set of formulas linked by informal text. We should choose the B variables such that each one of these formulas is represented by a variable. The B model the replaces the informal text between the formulas.

In any case, we should keep in mind that the goal is to define an acceptable behaviour for the device (acceptance criteria), not to write a physical model to describe accurately how the system really behaves. Finding an adequate level of modelling is very important to succeed. Often quite a few abstract variables and events are needed to cover all interesting formulas and cases concerning one physical variable.

As an example, we would not include the vehicle speed as a variable. Instead, we can use boolean variables representing formulas of the physical variables. We can for example define a variable, VehicleAtCruiseSpeed, which is true if the vehicle speed is within a certain defined interval of the set cruise speed.

4.3.3 Continuity and events

When using the B method, variables change value in events or procedures. If we are using events for changes in continuous expressions, we can get some events that can occur very frequently because of small changes in continuous variables. This is a problem, because we do not want models that generate a huge amount of events if nothing significant happens. Particularly this can make verification very complicated. Another potential problem with this modelling approach is that we can get quite many events for each variable. This means that we may end up with a model that is too complex compared to the physical reality described.

In general, we want to avoid these problems by defining appropriate variables and events, but sometimes this may constrain you so much that it is very difficult to model what you want. If this is the case, it may be better to use other modelling methods. The B method could not be expected to replace all other modelling methods, but it may be possible to combine them. This has not been studied, so we do not know how to combine the B method with continuous modelling.

We could also consider using the B method to model on a software level only. In such a model, it is easy to reason about discrete variables and events and map these to software. This approach would however give us a gap between the specification and the physical world, which means that it is difficult to determine if correct software behaviour also results in correct physical world behaviour. In this case study, we have tried to overcome this gap by choosing to specify with respect to the physical world.

4.4 Level of formalisation and capturing of requirements

The choice of variables, as discussed in previous section, also impacts the level of formalisation. What to formalise in a B model and what to leave in definitions and formulas made outside of B? Finding a good set of variables is important to get the right level of formalisation, avoiding both the case of too little formalisation and the case of too complex models.

Our analysis of the final specification, show that out of 36 original requirements, 19 were covered in the models and dictionary. Of the 17 uncovered requirements, 12 were deliberate choices of simplification in the case study, but 5 of these we would want to cover. These requirements were typically missed because the right scenario has not been considered. Of the 19 covered requirements, only 7 were considered sufficiently formalised, the remaining 12 requirements were mainly covered in the dictionary and need further formalisation. The requirements that need further formalisation are mainly covered by a few variables with complex definitions, for example `ObstaclePresent`, `VehicleTryKeepSpeed` and `VehicleTryKeepTimeGap`.

To avoid too much complexity in the user level model, one idea was to move some parts to another model, which could be used to refine and further formalise the complex variables. This model has not been completed, and we are not completely sure of how to fit it in the B refinement structure.

We can conclude that the need for further formalisation requires more refinements. Since the formalisation would require more variables and make the model more complex, the use of refinements is necessary to keep the user level model simple.

4.5 Verifying in the physical world

The obtained model consists of properties of the real world. To be able to verify this according to the "proven implementation" approach of the B method it is necessary to prove or in some way verify the formulas of the real world calculations obtained from the formulas of the B model. This should be possible in theory, but we do not know that because we have no concrete experience of this procedure, which is a new approach of B, very different from the more traditional use of verifying pieces of software against software level specifications.

Our experience from the case study indicates that obtaining the formulas of the real world can be quite difficult in some cases. We are for example interested in whether a vehicle behaves correctly or not with respect to its speed and acceleration. This is however very complicated due to many factors such as:

- Weight of the vehicle
- Available engine power
- Slope of the road
- Condition of the road
  - Is the vehicle acceleration reduced to avoid slipping?
  - Is vehicle deceleration reduced to avoid wheel-lock?
- Is the vehicle changing gears and thereby momentarily loses traction power?
- Other reasons for reduced performance

This results in very complicated formulas, particularly when we are reasoning about the vehicle behaviour, since we actually have to reverse engineer relevant control strategies to physical formulas. We risk the consequence of only obtaining a proven decomposition of physical formulas that are never written.

The decomposition model is leaving the vehicle behaviour and reasons about sub-system properties. However, some of these properties refer to the vehicle level behaviour. For example, the ACC sub-system shall output a torque requests that fulfils the vehicle behaviour required by the vehicle level model, given that other sub-systems work correctly. We are still dependent on the formulas discussed above.

With respect to verification, the approach of this case study only support formal verification of models. This verification ensures that requirements are not contradictory and that refined requirements fulfil top level requirements. This is a useful achievement, which to some degree ensures that the requirements are correctly stated. However, the implemented system must still be verified by traditional techniques such as testing. The formal models can be used as a guide for the testing, but since traditional manual techniques are still required, it is difficult to show any significant advantage. We have not studied this area in the case study, but it would be interesting to further investigate the verification process. Integration of other techniques should be possible, for example to automatically generate observers or test cases to improve the support for verification. Since this is not within the scope of the project, we still rely on traditional testing.

Without better support for verification, the usefulness of a formal user level model could be discussed. It may seem more appropriate to specify formally at the software level, because then it is more evident what is the correct behaviour and formal verification could be used. For example, the software would always try to maintain a certain speed even though it may not be possible with heavy load on a steep uphill slope. When we observe the software in this situation, we know that it is correct if it requests maximum engine torque. However, if we observe the vehicle decelerating in the same situation, it is not so easy to know if this is caused by incorrect software or external conditions.

For the user, the true physical behaviour matters, so the user level model is preferred before a traditional software specification. For a specification at software level, there is always a gap to the physical world and this gap is susceptible to errors. We must however also consider what we are able to verify and how efficiently we can perform verification. If the software verification can be formalised or automated, it may be more effective than traditional testing of the system. We can imagine that we want verification at both levels, the unanswered question is where and how to use formalisms to reduce the verification efforts.

4.6 Accuracy problems

4.6.1 Undefined values due to accuracy

One problem with reasoning about the real world is that we can never observe the real world without measurement errors. If we use simple discrete boolean variables to represent the fact that a certain formula holds between physical values, with a certain accuracy, which is the precise limit when the abstract discrete variable should change value?

Example:

`V` is a physical value (with infinite accuracy)

`Bx` is an abstract boolean used in the specification of a certain device `D` that should light a lamp when `V` is above a certain limit `l`.

We could define `Bx` as:

`Bx = TRUE if V >= l` (this formula being expressed in reals)

`Bx` = FALSE otherwise

This is not a good solution: the purpose of the B specification is to state that the lamp should light when `Bx` is `TRUE`. With the above definition, the lamp should light EXACTLY when `V` becomes greater than `1`, with infinite accuracy. No device can match such a specification.

Proposed solution: always define abstract variables with undefined states between defined ones, to reflect accuracy limits requested to the specified device.

For instance, let `2*p` be the requested accuracy for the specified device.

In the above example:

`Bx = TRUE if V >= 1+p` (this formula being expressed in reals)

`Bx` = FALSE if V <= l-p

`Bx` : BOOL if V in]l-p,l+p[

So if `V` is in `]1-p,1+p[`, the behaviour of the device is to be considered as correct whether the lamp lights or not.

Note: another solution would be to use 3 states variables, for instance:

`B3x = Above if V >= 1+p` (this formula being expressed in reals)

`B3x` = Below if V <= l-p

`B3x` = Middle if V in]l-p,l+p[

But it is more complicated.

4.6.2 Accuracy through refinement

Let us continue with the above example. We now suppose that we specify in a design level refinement that we have a measuring sub-device for `V`, that gives a discrete value `Vm`, and a lighting sub-device that will light if `Vm` is above a discrete limit lm.

As above we will not use NATURAL variables in the B refinement. Instead, we shall have a `Bm` boolean representing the fact that `Vm >= lm` (no accuracy required for discrete values). The definition of Bm will be:

`Bm = TRUE if Vm >= lm` (expressed in discrete values)

`Bm` = FALSE otherwise

As the refinement must be a specification for each sub-device of the device, the original `Bx` boolean variable should be kept in the refinement. Thus we will have both `Bx` and `Bm` in the refinement, with an invariant such as:

PropagationDone = TRUE => Bx = Bm

This is a specification for the measuring sub-device, it states that `Vm` should certainly be above `lm` when `V` is above `1+p`, and certainly below `Vm` when `V` is below `1-p`.

Will this cause accuracy problems in the way we manage to make `Bx` and `Bm` commute exactly together? No, in fact the precise point when `Bx` commutes is fixed by the way Bm changes, as long as Bx commutes somewhere in the]1-p,1+p[interval.

Bx is an abstract variable defined with a certain indeterminism that allows it to be linked easily to more concrete variables.

Note: it should be possible to get rid of the Bx variable and to write "PropagationDone = TRUE => Bx = Bm" as a gluing invariant. This does not change much, but then we need to keep track of the transitive gluing invariants in the refinement chain to translate the last refinement as an independent specification of each sub-device.

4.6.3 Observations in the model

This approach is implemented in the model. An example of the definition of accuracy is the variable ObstaclePresent, which has the following dependency to the distance:

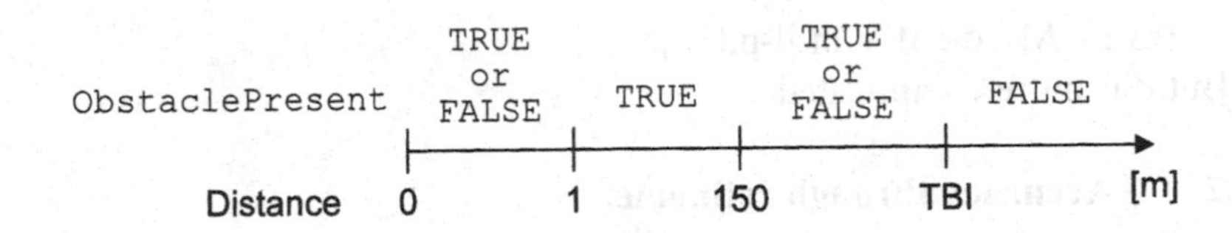

Figure -6. Definition of ObstaclePresent.11

In the refinement, the variable RadarSaysObstaclePresent represents the measurement from the radar. The variable RadarTransient represents the delay from a change in the environment until the radar has detected and reported this. The gluing invariant becomes RadarOn = TRUE & RadarTransient = FALSE => RadarSaysObstaclePresent = ObstaclePresent.

We can note that since the accuracy definition is entirely in the dictionary, it is not actually formalised, and consequently we do not prove any properties related to accuracy. This indicates the importance of the dictionary, whether a valid model represents a feasible system or not, is actually completely decided by the dictionary.

4.7 Meaning of model

We observed early that understanding the meaning of a B model is impossible if all variables and events are not properly defined. It is of course possible to read and understand the logical expressions in a B model, but understanding what this represents in the real world is a completely different story. Particularly it can be difficult to understand if a variable represents a physical or a logical property. It is essential to always give good definitions to all variables and events introduced.

In fact, the model has no meaning without proper definition of the variables and events used. These definitions are as important as the model itself. The formal model may look fine, but if you translate it to what it means in the real world, it can be something else than what you want. Particularly these problems can occur since the definitions of variables are informal and can be ambiguous and contradictory. To avoid this problem, variables should be carefully defined, using mathematical formulas if appropriate. Invariants should be used to avoid inconsistencies.

In the project, we have experienced some problems because the model has often progressed more rapidly than the dictionary. This can for example cause interpretation inconsistencies when you interpret the meaning of a variable a little different at different times and in different places in the model. The only way to avoid this is to clearly define the meaning of each variable, and use that definition each time the variable is used. Before you have the dictionary, it is also very difficult to say whether the meaning of the model is what you want or not. Even if the model is formally correct, it can represent the wrong requirements. All these problems are magnified when development is distributed, many people are involved, and when some people are method experts and other people are application experts.

4.8 Timing requirements

There is no explicit method for stating timing requirements in the B model. Timing requirements are represented by "delay" events, but the actual time of a deadline is only defined in the dictionary. The approach relies on using three events, one for the environment event, one for the system reaction event and one for the event when the specified time has passed. A variable is also used to represent that the delay is currently active, meaning that there is still some time left before a certain requirement must be fulfilled.

We try to explain this with a simplified example from our model. In our example we want to state the following requirement:

- When an obstacle is present, the vehicle shall adapt its speed to the obstacle within 200 ms from when the obstacle occurred.

Notice that this is a very simplified view, in reality we need to more carefully consider distance, speed difference, acceleration difference, cruise speed and whether cruise is active at all, but this simplified view is sufficient for this example. In our invariant we introduce the variable `NewObstacle`, to represent that there is an new obstacle present that occurred within 200 ms. We can also think of this variable as a delay activated by the obstacle and requiring reaction by the vehicle within a certain time. `VehicleTryKeepTimeGap` represents the fact that the vehicle adapts its speed to the obstacle. When an obstacle is present (`ObstaclePresent = TRUE`) either the vehicle has adapted its speed (`VehicleTryKeepTimeGap = TRUE`) or less than 200 ms has passed since the obstacle occurred (`NewObstacle = TRUE`).

```
INVARIANT
  ...
  (ObstaclePresent = TRUE
     => VehicleTryKeepTimeGap = TRUE or NewObstacle = TRUE)
  ...
```

Next, the environment event mainly marks the occurrence of something in the environment, using a variable to model that this event requires reaction from the system. In our example, the variable `NewObstacle` indicates that we are required to react to this event.

```
ObstacleAppears = SELECT
   ObstaclePresent = FALSE
THEN
   ObstaclePresent := TRUE &
   NewObstacle := TRUE
END;
```

The system reaction event states what are required from the system. This is typically stated as a change of the concerned variables such that the invariant is preserved when the delay has passed, though this event does not itself affect the delay.

```
VehicleManageObstacle = SELECT
    NewObstacle = TRUE
THEN
    ANY vtktg WHERE
        (vtktg : BOOL) &
        (ObstaclePresent = TRUE => vtktg = TRUE)
    THEN
        VehicleTryKeepTimeGap := vtktg
    END
END;
```

Finally, the timing event specifies that the time can only have passed if the invariant is preserved given that the delay is not active. In other words, the time can only pass if the required property holds. This type of event is very similar to the system reaction event, but the part from the invariant is now in the guard instead.

```
ObstacleDelay =
SELECT
    NewObstacle = TRUE &
    (ObstaclePresent = TRUE => VehicleTryKeepSpeed = TRUE)
THEN
    NewObstacle := FALSE
END;
```

We can conclude that it is possible to state timing requirements in this way in B. There are however several concerns with this approach.

The first problem is that stating the timing requirement gets more complicated the more variables that are involved in the requirement. Since we sometimes need several boolean variables representing one physical continuous variable, it is common to involve several variables in these requirements. In addition, stating timing requirements gets more complicated the more other requirements you have involving the same variables. Since we use the notion of change such that the invariant is preserved, we must include everything in the invariant that concerns the variables involved in the timing requirement. This may result in complicated events that represent something very simple, that a certain time has passed. This is not so convenient, neither for writers, nor for readers of the specification. The solution to this problem is to use a definition of the concerned expressions in the invariant. This definition can then be referred in the invariant, the system reaction event and the timing event. Notice that the events need a parameterized definition.

The second problem is that this approach has not been used down to the design or implementation levels, excepted on small examples. So there is a risk concerning the difficulty of proofs involved. Since we have no experience from refining such a specification to an implementation, we do not know how difficult this would be, but we the think that such proofs would be extremely difficult to perform using the B method.

To formalise the statement "time cannot pass unless condition A holds", is a way to formalise timing requirements, but it does not provide much support for verification of the timing requirements. Stating that something bad cannot happen is obviously not sufficient to prevent it from actually happening. At this level of abstraction, formalising such a statement gives no apparent advantage. Only if we can reach a design, many refinement steps ahead, we could gain some value by proving that the design really fulfils the abstract specification. We do not even know if this is feasible, since we have not managed to reach this level of refinement. Again, we need to rely on traditional testing to verify these requirements.

Some studies are performed to see how the RAVEN model checker can be used to support verification of these requirements. This study is not finished, but the main advantage is probably abilities to verify that there are no dead-locks (or live-locks) in the system, rather than verifying actual timing requirements.

4.9 Tool experience

4.9.1 General impressions

AtelierB was experienced as a stable tool, that is easy too work with most of the time. The most difficult and most important activity in the tool is the interactive proof. Unfortunately this part is not particularly intuitive and user friendly. Information is presented in a way that often makes it difficult to read, and you need to search for the essential information. The available help for commands is not sufficient. There is hardly any guidance or hints on suitable proof strategies. If there is an error in the specification, you have to figure out the problem very much yourself. The tool has for example hardly any support for finding and showing counterexamples. We have seen some promising demonstrations of ideas of improvements addressing these issues, though these are not yet available in tools.

Regarding the complementary tools being developed in the project, we have only really tried the U2B tool. Since this tool has been a prototype under development during the project, it has been somewhat difficult to use.

At the end of the project, most translations are becoming more stable and well defined, so we have eventually managed to use it after some guidance from Southampton. The tool as such is simple to use, but you need good documentation to know how to make models in UML that can be translated to valid B models. Our impression now is that the tool is sufficient mature and complete to be useful. Once the documentation is up-to-date with the tool, anyone with some UML and B knowledge should be able to utilise the tool. However, to perform the proofs, you still need to have quite deep knowledge in B. And performing the proofs is quite an essential task of the formalisation of a model.

4.9.2 Working several persons in the same project

To use AtelierB in a multi-user environment it is recommended to use a version control tool. Working without a version control tool is very risky, since several persons then can edit the same file at the same time, which can result in loss of work. In this project we used CVS (Concurrent Versions System), which is a commonly used version control tool in UNIX environments. This works without any problem for the B model and dictionary source files.

We also want to share the proof files, since we do want to each perform proofs that someone else has already done and we do not want to lose performed proofs when someone else updates the model. Sharing the proofs is a bit more complicated though, using the same approach as for the source files would probably not work.

We have used two other approaches to save and share proofs. The first approach is to use a shared project for doing all proofs. In this way, all proofs are kept in one place and are never manipulated by the version control tool. It is perfectly safe to work concurrently in the same project when doing proofs, since the source files are never modified. The disadvantage is of course that whenever an error in the model is detected during the proof process, it is necessary to change to a private project, modify the model, check in the model, change to the shared project and check out the model. In practice, this is not a problem, since we work in private projects until the model is sufficiently mature, only then is it necessary to perform any remaining difficult proofs in the shared project. As a second approach, we archive the project, including all proofs, regularly and stores a history of these archives in our version control tool.

4.10 Dealing with complexity

We have encountered several difficulties related to complexity. The first observation is a risk to put complex things in the dictionary only, to avoid formalising them. In fact, a large amount of the original requirements that are covered are only covered in the dictionary. These requirements will thus not be included in any proofs. An example is two of the key variables, `VehicleTryKeepSpeed` and `VehicleTryKeepTimeGap`, which are essential for the definition of an ACC. These variables are completely defined, using complex formulas, outside of B. We can of course imagine that these can be moved from the dictionary to the model in a refinement further on, or split into more detailed formalised models in additional higher levels.

Another observation is that the effort of adding to the model grows, at least linearly, with the size of the model. For each added variable, we need to consider how each existing event affect this variable. For each new event, we need to consider which of the existing variables it affects. We have already reached a point where the work with the model is getting difficult, still there is lot of simplifications made to the model. We do not know if it would be manageable to make a sufficient complete model, which would probably be many times larger than our present model.

Problems with complexity of course occur no matter what approach you use if your project develops a complex system. Using formalisms, you are forced to deal with this quite early and make detailed models. Using traditional methods, there is a tendency to abstract away certain problems in the beginning and have them appear later in the project, typically in the implementation stage. This is exactly what we want to avoid, so it can be expected to have to deal with complexity and to spend more time in an earlier stage.

Using a formal method, complexity means harder proofs. We can see a tendency to modify your model in order to simplify proofs. A model should of course not be unnecessary complex, but simplifying models can also mean to move some problems from the formal model to the informal dictionary. Assuming we want to use formal methods to prove that we have solved the complex problems correctly, they also need to be formalised. Eventually then, we need to make more complex models and the proofs will be more difficult. We can see already from this case study, that some proofs get complicated due to complexity. The problem is generally not that the proofs are particularly difficult, but that they are large, involving many variables and many hypothesis formulas. We believe that in the case study, we have only seen the start of this problem and that this would grow significantly with the project if we would continue using formal methods.

New strategies to deal with complex proofs may be needed. Particularly, this applies to projects developing something new and complex, when you often make many attempts and changes a lot before you are satisfied. Efforts for changing models and redoing proofs could then be a limiting factor. This seems to be a particular problem when modelling in UML, because this adds proof complexity to the B model.

4.11 Model confidence

Using formal methods, you can prove that a design, or even an implementation, is correct with respect to a specification. However, you cannot guarantee that the original specification is correct, meaning that it is not generally possible to detect missing or incorrectly stated requirements. This does not imply that formal methods should only be used to address the first of these two problems. In fact, we believe that most errors are made when a requirement is first stated. If a requirement is properly stated, it is rarely disrespected by an implementation. Most problems are due requirements that are unclear, ambiguous, contradictory, incorrect or missing. The nature of a formal language should help us avoiding unclear and ambiguous requirements. Consistency checks can be used to find contradictory requirements. Incorrect or missing requirements are more difficult, we cannot get any guarantees, but the method may help us by encouraging us using sound practices. It makes sense to reason about our confidence in the specification, particularly compared to the confidence of an informal specification.

We know that we only prove that the model preserves the invariant and that each refinement fulfils the abstract specifications. The tool will help us discover contradictions and certain errors in the model. However, if we forget to put something in the invariant, there is no way to detect this. In some cases, an incomplete invariant will result in difficult proofs, so we may discover this, but we have no guarantees. In fact, no matter what method we use, we can never know if we have forgotten one requirement. What is essential for the B method is to be sure that all requirements we know are really stated in the model. The same applies to stating the right requirements, no method will detect whether our model really represents the requirements we want. These types of errors must be detected manually by carefully reviewing the model.

Our experience from the case study is that it is difficult to ensure that a B model really represents the requirements you want. Proving a model is not sufficient; you can still have a model with missing or incorrect requirements. During the most part of the project, we have felt quite unconfident on the correctness of the model. Of course, we can expect that it will take some time before a correct model is obtained, but we have never really stopped discovering and fixing errors in the model. We cannot explain the reason for this. Maybe the case is so complex that it takes so long time to reach a correct specification, or our modelling has not been so good.

Some errors that we have discovered are incorrect modelling of requirements, meaning that our model does not represent the requirements that we want. We have also discovered some problems with consistency. Although the B model itself is guaranteed to be consistent, the dictionary is not. In our case, the dictionary contains a lot of information, and is essential for the meaning of the specification. We have identified inconsistencies in the dictionary, and between the dictionary and the B model, which means that there are inconsistency problems when the model is interpreted even though the B model is logically correct.

Another tricky problem is incorrect constraints on the environment. We have at least one example of missing events that means that the model incorrectly prevents changes in variables that should be possible. The identified case is considered of minor importance, and was one of the problems that were discovered late, so it has not been fixed. This was also an example of a typical error that is not detected by proofs in B, but it was detected using an animator to validate the specified behaviour. Even if this particular error would probably not result in an incorrect system, verification would be prevented since it is not possible to decide whether the system is correct or not if the environment behaves in an undefined way.

A similar problem is events with guards that are too strong. This could probably be handled correctly in the implementation level in procedural B, but it is not apparent in the event B specification. When something that cannot happen according to the specification happens in reality, the result is often something bad. Again, it is necessary to carefully consider constraints on the environment to ensure that the intended behaviour will be achieved in reality.

Discovering and fixing errors also means redoing some proofs. In our case, it resulted in spending quite some time in several proof iterations. Particularly the refinement was tricky to prove, and changes in the invariant meant that all proofs have to be redone.

5. CONCLUSIONS

The specification as such is an interesting result of the case study. It is an example of an abstract model specifying a system. Usually, this kind of model is rarely found at all. The UML version of the model is of particular interest since it should be accessible for a wider range of audience.

The work with the modelling has been an experiment of how modelling methods. The approach has often been invented in the case study and tried for the first time. The case study has also involved a lot of learning of methods. Due to these circumstances, extra time has been needed for many unforeseen problems.

Some of the difficulties we have experienced are likely related to the nature of the case study. We think that B is not equally suitable for all kinds of modelling, making some types of systems difficult to model using B. Our case study may represent a rather difficult system to model in B, for example due to a semantic gap between what we want to model and what we can express in B. For another type of system, we could imagine that these difficulties could be significantly smaller.

Use of formalisms has given us some support in creating a better model, mainly by ensuring consistency between requirements. However, we can also conclude that we cannot rely on formalisms for many types of errors, for example missing or incorrectly stated requirements.

Even though this was to a large degree a learning project, we can conclude that the use of formal methods require a lot of extra time. Creating, understanding and proving a formal model is more complicated than specifying the same thing informally. It seems likely that spending the same time on an informal specification would result in an equally good specification.

In the case study, we have only addressed the quality and verification of a specification. We have however no extra support for the verification of a system, but we would still have to rely on manual verification.

Using B requires a lot of education and practice to master. Maybe it is only wise to use it at all if you are an expert. Using the UML approach makes it easier to understand and create models, but you still need to be an expert to do the proofs. Adopting a typical engineer to using B would be very difficult unless you already have a team with B experts as support.

Given the few apparent benefits, it is difficult to motivate the efforts from using B. To use only for specifications, it should be much easier to use. The UML modelling front-end is a part of the solution, but the proof engine should then be much more capable of automatic proofs and on pointing out errors. If the method would also give support for verification of a system, then it would be easier to motivate the extra efforts with modelling and proofing.

Chapter 12

THE ADAPTIVE CRUISE CONTROLLER CASE STUDY

Visualisation, Validation, and Temporal Verification

Alexander Krupp[1], Ola Lundkvist[2], Tim Schattkowsky[1] and Colin Snook[3]
[1]Paderborn University; [2]Volvo Technology Corporation; [3]University of Southampton

Abstract: We present the adaptive cruise controller case study for B modelling and the model checking by RAVEN. Individual translations of B operations, data types, and invariants to the RAVEN Input Language are presented by the example of the case study.

Key words: aumotive systems, formal refinement, model checking, UML

1. INTRODUCTION

We have written a B specification for an adaptive cruise control following the lines of Chapter 5 and 6, respectively. The specification was verified as internally consistent but was difficult to capture even to those involved in its creation. Thus, an additional validation of the model was needed to convince the stakeholders that the model had the required behaviour. We therefore manually translated the model into a UML-B version in order to make it easier to understand and reason about. We automatically translated the UML-B model into a new B version using U2B and animated it using the ProB model checker/animator [5]. The animation did not reveal any problems with the model but substantially improved confidence and understanding in its behaviour.

In the process, some temporal requirements were deduced that had not been covered by the previous consistency verification. In order to check those properties, the B code produced by U2B was translated into RIL and verified by the RAVEN model checker [4]. For that, we manually translated the B code, which was generated by the UML2B tool to the RAVEN Input

J. Mermet (ed.), UML-B Specification for Proven Embedded System Design, 199–210.
© 2004 *Kluwer Academic Publishers. Printed in the Netherlands.*

Language (RIL) and verified the model and the specification through additional model checking. No problems were found during the temporal verification. As the structure of the existing B specification of the cruise controller is quite different from the approach, which was undertaken in Chapter 10 and in [2,3], we took a different approach in the translation to RIL. We have performed that complementary model checking of temporal properties since model checking additionally allows for reasoning about state sequences, which is expressible, but rather difficult to prove in B. The main advantages of the RAVEN model checker are:

- counter example generation for invalid sequences of operations,
- data analysis (min/max values of variables), and
- time analysis (min/max time of specific state configurations).

However, due to the state explosion problem, the main general drawback of model checking is of course the rather limited state space to be explored. Usually, abstraction techniques have to be applied to reduce the state space. As B offers a verification path from an abstract system towards an implementation, it seems natural to apply model checking at an abstract level in B, when a state space can be kept small.

The remainder of this chapter is structured as follows. The next two sections outline the translation of B operations, data types, and invariants to the RAVEN Input Language (RIL) and their visualisation as UML-B. Thereafter, verification results are presented.

2. VISUALISATION OF THE MODEL IN UML-B

To obtain an adequate visual UML-B presentation of the B model, we firstly examine B for indications of modelling of a set of instances. Typically, if several B operations have a parameter with the same type and this type is used in the domain of some function type variables, they can be combined to classes with class diagrams for representation. However, in the B model, no sets of instances are modelled in this way. Our second strategy was to look for a variable that is used as a guard and then changed to a new value by several operations. The life cycle of such a variable can be usefully represented as a state diagram with the operations at transitions. At first sight, there appeared to be no such variable, however, the invariant indicated that a group of Boolean variables that were used as guards in operations contained exclusions, e.g., `var1=F=>var2..n=F`. Such a group of variables was found to collectively represent a state variable. By replacing the group with a single variable representing the state, a better visual representation from a different perspective was achieved. We discovered no disadvantages of combining these Booleans although we did not attempt a refinement. This representation generated a collation of related operations that all transitions the system between its main operational conditions such as `cruiseBecomesNotAllowed`, `cruiseOff` etc.

Similarly, a variable representing the state of the closest obstacle was identified resulting in a collation of operations related to the obstacle status. We decided on a natural segregation of the model into two classes as given in Figure 1.

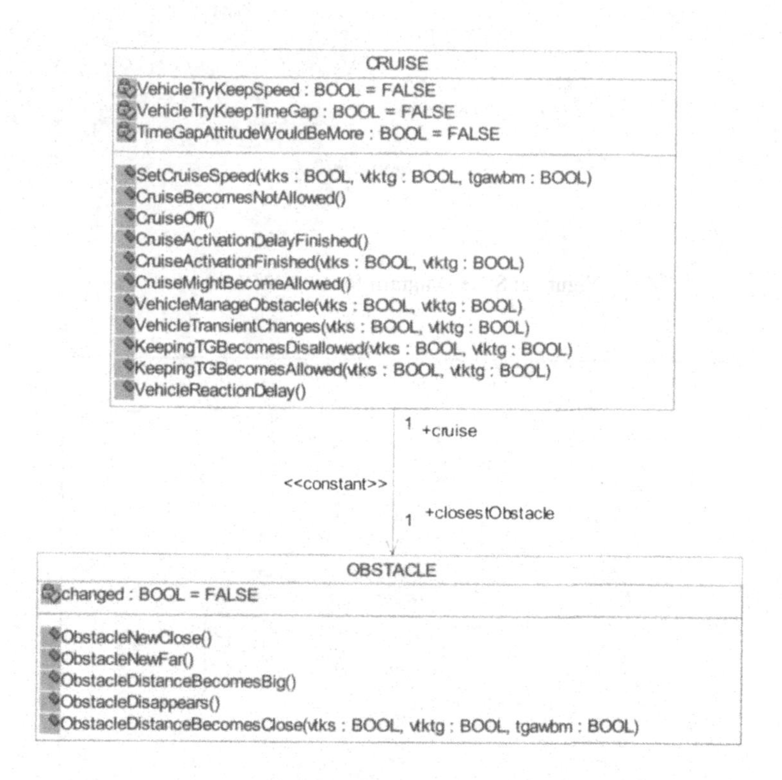

Figure 1: Class Diagram for the Cruise Controller

The classes are `CRUISE` and `OBSTACLE`. Note, that the case study, just has a single instance of each class. This enabled the use of an association to represent closestObstacle (since associations are links between instances), which either is an unnecessary complication compared to the original B model, or is necessary to create valid invariants for UML-B model. For each class, the behaviour is defined as a state diagram. Figure 2 gives the state diagram from class `OBSTACLE` and Figure 3 shows the corresponding B code with invariants.

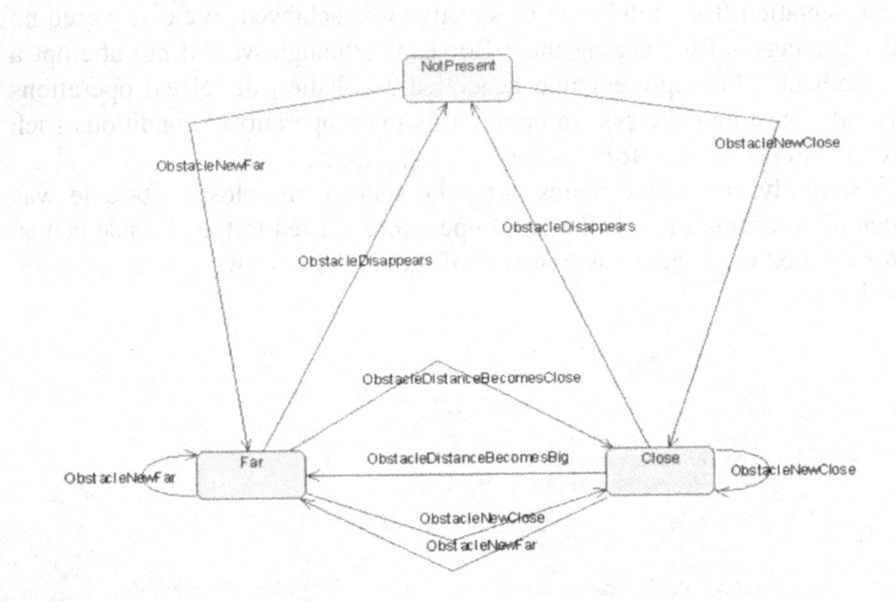

Figure 2: State Diagram for OBSTACLE

```
CARDINALITY 1
DEFINITIONS
CruiseReallyActive ==
      (closestObstacle.changed = FALSE &
       cruise_state/=CruiseActivationInProgress);
AccLaws(od, vtks, vtktg, tgawbm) ==
      /* do something law */
      (vtks = TRUE or vtktg = TRUE) &
      /* keep speed if far or no obstacle */
      (od :{ Far,NotPresent} => vtks = TRUE) &
      /* keep time gap otherwise, if not too fast */
      ((od = Close & tgawbm = FALSE) => vtktg = TRUE) &
      /* keep speed, if time gap
              keeping would be too fast */
      ((od = Close & tgawbm = TRUE) => vtks = TRUE)
INVARIANT
  (cruise_state:{CruiseNotAllowed, CruiseAllowed} =>
      VehicleTryKeepSpeed=FALSE) &
  (cruise_state:{CruiseNotAllowed, CruiseAllowed} =>
      VehicleTryKeepTimeGap=FALSE) &
  (cruise_state:{CruiseNotAllowed, CruiseAllowed} =>
      closestObstacle.changed=FALSE) &
  (cruise_state:{CruiseNotAllowed, CruiseAllowed} =>
      TimeGapAttitudeWouldBeMore=FALSE) &
  (closestObstacle.obstacle_state=NotPresent =>
      VehicleTryKeepTimeGap=FALSE) &
  (closestObstacle.obstacle_state:{NotPresent,Far} =>
      TimeGapAttitudeWouldBeMore=FALSE) &
  (cruise_state:{CruiseActive,CruiseActivationInProgress} &
      CruiseReallyActive =>
      (AccLaws(closestObstacle.obstacle_state,
      VehicleTryKeepSpeed,
      VehicleTryKeepTimeGap, TimeGapAttitudeWouldBeMore)
      )
  )
INITIALISATION
closestObstacle:=thisOBSTACLE
```

Figure 3: UML-B Invariants

3. TRANSLATION OF B OPERATIONS AND DATA TYPES

In contrast to the automatic approach of Chapter 10 and 14, we take a pragmatic approach of a direct translation from the B model to the RAVEN Input Language (RIL) here. We outline how to translate the B model of Volvo's Adaptive Cruise Controller, generated from the UML2B tool to the RIL. For further information on that model and on the case study, please consult Chapter 10 or [3,4].

When translating from B to RIL directly, we have to start with the examination of the state space of the B model, i.e., we examine the set of B variables as given in the following enumeration.

```
B code
VARIABLES
  obstacle_state,
  changed,
  cruise_state,
  VehicleTryKeepSpeed,
  VehicleTryKeepTimeGap,
  TimeGapAttitudeWouldBeMore,
  NumberOfSetCruise
```

Additionally, the type of each variable from the B invariant has to be considered.

```
B code
obstacle_state: OBSTACLE_STATE &
changed: BOOL &
cruise_state: CRUISE_STATE &
VehicleTryKeepSpeed: BOOL &
VehicleTryKeepTimeGap: BOOL &
TimeGapAttitudeWouldBeMore: BOOL &
NumberOfSetCruise: NATURAL
```

The four Boolean variables in the B code translate to RIL easily. The enumerated set variables `OBSTACLE_STATE` and `CRUISE_STATE`, have rather small associated sets specified in the SETS clause as given hereafter.

```
——————————— B code ———————————
SETS
  OBSTACLE_STATE = {NotPresent, Close, Far};
  CRUISE_STATE   = {CruiseNotAllowed,
                    CruiseAllowed,
                    CruiseActive,
                    CruiseActivationInProgress};
```

Since the unrestricted natural variable `NumberOfSetCruise` would cause a state explosion with a BDD model checker, we additionally have to restrict the value of `NumberOfSetCruise` for model checking to a smaller interval, so that the translation the complete set of variables to RIL is as follows.

```
——————————— RIL code ———————————
MODULE Cruise
SIGNAL
  obstacle_state: NotPresent Close Far
  changed: BOOL
  cruise_state: {CruiseNotAllowed,
                 CruiseAllowed,
                 CruiseActive,
                 CruiseActivationInProgress}
  VehicleTryKeepSpeed: BOOL
  VehicleTryKeepTimeGap: BOOL
  TimeGapAttitudeWouldBeMore: BOOL
  NumberOfSetCruise: RANGE[0,4]
```

We can then continue with translation of the B operations starting with a set of B operations, which may be triggered any time their preconditions are valid. As each of the transitions executes due to a different set of source states, we cannot declare a specific set of variables as source states for the RAVEN transitions. Instead, we simply write *TRUE* as the source state for the transitions. The execution of each transition is then dependent on its *guard*. Generic transitions in RIL correspond to the following patterns:

```
|- [source state] -- [guard] --> [target state]
```

Consider the operation `ObstacleDisappears` as an example for translations of B operations to RIL.

```
------------------------- B code -------------------------
ObstacleDisappears =
  BEGIN
     SELECT obstacle_state=Close
     THEN obstacle_state:=NotPresent
     WHEN obstacle_state=Far
     THEN obstacle_state:=NotPresent
  END ||
  IF cruise_state : {CruiseActive,CruiseActivationInProgress}
  THEN
     changed := TRUE ||
     TimeGapAttitudeWouldBeMore := FALSE ||
     VehicleTryKeepTimeGap := FALSE
  END
```

The operation has two simultaneous substitutions. In RIL, we create one transition for each combination of choices in the simultaneous substitutions.

```
------------------------- RIL code -------------------------
// ObstacleDisappears
|- TRUE -- (obstacle_state=NotPresent) &
           !((cruise_state=CruiseActive) |
             (cruise_state=CruiseActivationInProgress))
       --> obstacle_state:=NotPresent
|- TRUE -- (obstacle_state=Close) &
           !((cruise_state=CruiseActive) |
            (cruise_state=CruiseActivationInProgress))
       --> obstacle_state:=NotPresent
|- TRUE -- (obstacle_state=Far) &
           !((cruise_state=CruiseActive) |
             (cruise_state=CruiseActivationInProgress))
       --> obstacle_state:=NotPresent
|- TRUE -- (obstacle_state=NotPresent) &
           ((cruise_state=CruiseActive) |
            (cruise_state=CruiseActivationInProgress))
       --> obstacle_state:=NotPresent; changed:=TRUE;
           VehicleTryKeepTimeGap:=FALSE;
           TimeGapAttitudeWouldBeMore:=FALSE
|- TRUE -- (obstacle_state=Close) &
           ((cruise_state=CruiseActive) |
            (cruise_state=CruiseActivationInProgress))
       --> obstacle_state:=NotPresent; changed:=TRUE;
           VehicleTryKeepTimeGap:=FALSE;
           TimeGapAttitudeWouldBeMore:=FALSE
|- TRUE -- (obstacle_state=Far) &
           ((cruise_state=CruiseActive) |
            (cruise_state=CruiseActivationInProgress))
       --> obstacle_state:=NotPresent; changed:=TRUE;
           VehicleTryKeepTimeGap:=FALSE;
           TimeGapAttitudeWouldBeMore:=FALSE
```

A scheme for translation of simultaneous choice substitutions can be formulated in generalized substitution notation as follows.

```
(g1 ⇒ [S1]) [] (g2 ⇒ [S2]) [] . . . [] (gn ⇒ [Sn]) ||
(h1 ⇒ [T1]) [] (h2 ⇒ [T2]) [] . . . [] (hn ⇒ [Tn])
```

This is presented in RIL by creating one transition for each combination of simultaneous guards and their associated combination of substitutions as given in the following code. Note here, that substitutions need to be simple assignments, though.

```
——————— RIL code ———————
|- TRUE -- g1 & h1 --> S1; T1
|- TRUE -- g1 & h2 --> S1; T2
...
|- TRUE -- g1 & hn --> S1; Tn
|- TRUE -- g2 & h1 --> S2; T1
|- TRUE -- g2 & h2 --> S2; T2
...
|- TRUE -- g2 & hn --> S2; Tn
...
|- TRUE -- gn & h1 --> Sn; T1
|- TRUE -- gn & h2 --> Sn; T2
...
|- TRUE -- gn & hn --> Sn; Tn
```

As a more complex example, consider the following B operation. It takes two Boolean parameters. Additionally, it uses a B definition called `AccLaws`, which is expanded and adapted to RIL. The parameters `vtks`, and `vtktg` inflict a more substantial change on the generated RIL. As the operation precondition accepts any combination of parameter values at any time, we need to emulate the operation invocation by a separate module in RIL. This module contains representations of all variables used as parameters. With each execution step it sets the variables to an arbitrary configuration. That means, that each execution step in the RAVEN execution sequence offers all possible parameter input valuations. The B operation also uses a nested choice construct, which requires a slightly different translation than before.

```
B code
ObstacleDistanceBecomesBig (vtks,vtktg) =
PRE
  vtks:BOOL &
  vtktg:BOOL
THEN
  SELECT
     (CruiseReallyActive => AccLaws(Far, vtks, vtktg, FALSE))
  THEN
     SELECT obstacle_state=Close
     THEN obstacle_state:=Far
     END ||
     TimeGapAttitudeWouldBeMore := FALSE ||
     IF
        cruise_state: {CruiseActive, CruiseActivationInProgress}
      THEN
        VehicleTryKeepTimeGap := vtktg ||
        VehicleTryKeepSpeed := vtks
     END
  END
END
```

It can be observed in the B code, that there is a first `SELECT` statement enclosing the remainder of the operation. The guard of this `SELECT` is therefore in every generated RIL transition. The enclosed part is translated as described earlier. The additional guards generated from it are combined with the guard of the enclosing `SELECT` statement.[1]

```
RIL code
// ObstacleDistanceBecomesBig
 |- TRUE -- (CruiseReallyActive ->
             ((vtks|vtktg) & vtks & TRUE & TRUE)) &
             (obstacle_state=Close) &
             ((cruise_state=CruiseActive) |
             (cruise_state=CruiseActivationInProgress))
        --> obstacle_state:=Far;
             TimeGapAttitudeWouldBeMore:=FALSE;
             VehicleTryKeepTimeGap:=vtktg;
             VehicleTryKeepSpeed:=vtks
 |- TRUE -- (CruiseReallyActive ->
             ((vtks|vtktg) & vtks & TRUE & TRUE)) &
              (obstacle_state=Close) &
              !((cruise_state=CruiseActive) |
              (cruise_state=CruiseActivationInProgress))
        --> obstacle_state:=Far;
             TimeGapAttitudeWouldBeMore:=FALSE
```

[1] Note that, corresponding to the B book [1], a `SELECT` substitution can become *infeasible* if it does not contain an `ELSE` part and it is possible that none of it guards hold. When a simultaneous substitution contains one infeasible substitution, then the whole substitution becomes infeasible. Our translation implicitly generates transitions for *feasible* cases only.

4. TRANSFORMATION OF B INVARIANTS

This section sketches how parts of B invariants can be transformed to RIL specifications for verification by model checking. A translatable invariant has to comply to the following rules.

1. It may only refer to identifiers which conform to RIL types: `INT`, `NAT`, `BOOL`, and enumerated `SETS`.
   It may only use unary and binary operators supported by CCTL:
   $\leftarrow$, $\wedge\leftarrow\leftarrow\vee$, $\Rightarrow\leftarrow\Leftrightarrow$, , , , ,

Each compliant invariant can be translated to RIL. The B operators have to be replaced by their RIL equivalents, of course, and some additional brackets may have to be introduced due to different operator precedence. The invariant is then prefixed in RIL with the temporal operator.

The following B code shows a fragment of the B invariant, which can be translated to RIL as two separate CCTL formulae.

```
B code
... &
(cruise state = CruiseNotAllowed =>
  (VehicleTryKeepSpeed = FALSE &
  VehicleTryKeepTimeGap = FALSE &
  changed = FALSE &
  TimeGapAttitudeWouldBeMore = FALSE &
  NumberOfSetCruise = 0)) &
(cruise state = CruiseAllowed =>
  (VehicleTryKeepSpeed = FALSE &
   VehicleTryKeepTimeGap = FALSE &
   changed = FALSE &
   TimeGapAttitudeWouldBeMore = FALSE &
   NumberOfSetCruise = 0))
& ...
```

The next code gives the translated invariants in combination with additional checks. The translated B invariants are contained in the two specifications *binv3* and *binv4*. In CTL, such expressions are called *safety conditions*. Specification *l1* is a reachability condition. It states, that it is always possible that the state with identifier `changed` can be eventually reached.

```
------------------------------ RIL code ------------------------------
SPEC
...
binv3  := AG
          ((Cruise.cruise_state =
            Cruise.CruiseNotAllowed) ->
              (!Cruise.VehicleTryKeepSpeed &
               !Cruise.VehicleTryKeepTimeGap &
               !Cruise.changed &
               !Cruise.TimeGapAttitudeWouldBeMore &
               (Cruise.NumberOfSetCruise = 0)))

binv4  := AG
         ((Cruise.cruise_state =
           Cruise.CruiseAllowed)->
             (!Cruise.VehicleTryKeepSpeed &
              !Cruise.VehicleTryKeepTimeGap &
              !Cruise.changed &
              !Cruise.TimeGapAttitudeWouldBeMore &
              (Cruise.NumberOfSetCruise = 0))

l1  := AG EF Cruise.changed
...
```

5. RESULTS

The verification with RAVEN has shown, that all invariants translated from B are correct. The overall execution time of the translated RIL example for RAVEN model checking and analysis was less than a second on an Intel P4, 2.2GHz, 1GB RAM. Memory allocation of RAVEN was approximately 3.9MB.

In addition to checking restricted B invariants, RAVEN is able to check constraints, which involve state sequences. Among those are additional qualitative properties like safety, reachability, response, inevitability, and liveness constraints for state-based verification. Especially with Event-B it is useful to be able to check for possible sequences of states. In addition, RAVEN performs quantitative and timing analysis. It allows to determine minimum or maximum reaction times, and to check for minimum and maximum values of integers within a certain set of states. Some examples of quantative data analysis can be found in Chapter 14 of this book.

However, our focus here was on a specification flow from B generated by UML2B to RAVEN in contrast to Chapter 10 and 14, where we investigate B patterns to verify the behavior of finite state machines. The feasibility of this approach was shown under specific restrictions on B.

Nevertheless, RAVEN's analysis capabilities give a good example that it makes sense to use B in combination with other formal verification means like model checking, which gives complementary verification and additional confidence in the model under design.

REFERENCES

1. J.R. Abrial. The B-Book. Cambridge University Press, 1996.
2. A. Krupp, W. Mueller. Deliverable D4.3.1: Specification of Real-Time Properties. Technical Report, IST - Project PUSSEE, December 2002.
3. A. Krupp, W. Mueller. Deliverable D4.3.2: Refinement and Verification of Real-Time Properties. Technical Report, IST - Project PUSSEE, April 2003
4. J. Ruf. RAVEN: Real-Time Analyzing and Verification Environment. J.UCS, Springer, Heidelberg, 2001.
5. P.J. Turner, M.A. Leuschel. ProB V1.06 User Manual. Declarative Systems and Software Engineering, University of Southampton, UK, 2004.
6. J. Zandin. Non-Operational, Temporal Specification Using the B Method – A Cruise Controller Case Study. Master's Thesis, Chalmers University of Technology and Gothenburg University, 1999.

Chapter 13

FORMAL MODELLING OF ELECTRONIC CIRCUITS USING EVENT-B

Case Study: SAE J1708 Serial Communication Link

Yann Zimmermann[1,2], Stefan Hallerstede[1], Dominique Cansell[2]
[1]KeesDA, Centre Equation, 2 Avenue de Vignate, 38610 Gières; [2]LORIA, BP 239, 54506 Vandoeuvre-Lès-Nancy Cedex

1. INTRODUCTION

This chapter presents a study of the SAE J1708 Serial Communication link described in [1]. The study is carried out in Event-B, an extension of the B method. The system is implemented and decomposed using step-wise refinement. We present how to derive with this method a cycle-accurate hardware model. The model of the communication link system is composed of an arbitrary, finite, number of identical components that run concurrently. The model contains synchronization of these components required to control access to the communication link. At the end of the refinement we obtain an implementable model of the components which is translated into VHDL. The generated VHDL design is synthesizable, meaning that the implementable B model is synthesizable as well.

The system is first described in an abstract way. Then, step by step, it is described in more detail by means of refinement. The refinement process permits to develop the system step by step. Each step introduces an aspect of the system. It allows us to cope better with the complexity of a model than if a model was obtained at once. Refinement adds detail to a model, distributes complexity of the system and of its proof. It also makes it easier to explain, and communicate to achieve a step by step validation of the model. The last refinement has to be close to an implementation. Particularly, we aim to obtain a model of the system in which the description of the behavior of each

J. Mermet (ed.), UML-B Specification for Proven Embedded System Design, 211–226.
© 2004 *Kluwer Academic Publishers. Printed in the Netherlands.*

controller only refers to its local state. We say that the components are independent.

The system is described in terms of a collection of events. Events permit to model concurrency. When more than one event may occur at any one time, the choice of what event occurs is non-deterministic. The independence of components ensures concurrency between the controllers because it does not impose any order on their execution. For this study we have made the assumption that a component writing to the bus immediately stops writing to it, if it registers a conflict in correspondence with ([1, 5.2.2]: "or sooner if possible"). In this case, bus contention is always resolved. We refer to this idealized model as the "ideal model" and to the other model as the "probabilistic model". From the initial specification of the ideal model we have developed a bit-level implementation of the bus-access controller. We prove that this model is safe, in that bus contention is always resolved (without delay). The advantage of this approach is that we will understand the implementation of the (ideal) controller before taking into account the technical difficulties involved in the probabilistic model. Furthermore, we consider the ideal model to be of interest by itself since it is implementable and may be used in future systems (where all components comply with the idealistic requirement above).

2. GLOSSARY

The glossary defines the concepts that are modeled formally. It is an important step in analyzing the intent of a specification. While creating the glossary we remove inconsistencies in the use of technical terms in the standards model. These inconsistencies usually arrive in natural language specifications. Formalizing such a specification forces us to define all terms precisely, thus revealing inconsistencies.

2.1 The system

The system consists of a network of components which communicate via a bus. All components are connected to the bus by a local controller. They may send (or receive) messages to (from) the bus.

Network: The network interconnects all components of the system via a global bus.

Node: Components are connected to the bus by a node. The node is the controller part of a component. A node is a receiver or a transceiver. A transceiver is a transmitter or a receiver.

Etc…

2.2 Contention resolution

Our aim is to model the contention resolution between transmitters. The reception of a message is not modeled. Data is sent on the bus bit by bit. Each bit has a duration. It is not necessary to model this duration in our model.

Bit time: The bit time is the duration or period of one unit (bit) of information.

Character time: The character time is the duration of one character being transmitted.

Etc…

The glossary has about 3 pages. The standards document has 10 pages. The clarification of the terms used in the standard is important to arrive at a meaningful B model.

3. EVENT-B

Event B is an extension of the B method [2, 3]. This method is based on two concepts: proof and refinement. Beginning with a simple abstract machine, complex systems are refined step by step, increasing the level of detail at each step. Each refinement is justified by a formal proof. The initial abstract machine establishes main properties of the system. The number of established and proved properties increases with each refinement. When a sequence of refinements is completed, the system is described at a chosen level of detail, and every property introduced during the chain of refinements is satisfied.

4. SPECIFICATION AND IMPLEMENTATION OF THE SERIAL LINK

In the following we develop the specification of the standard in five refinement steps. Due to space limitations we only present the first and the last step though. The overall development takes 17 refinements. The remaining 12 steps are used to implement the system, more precisely the bus controllers. The proof effort is summed up in Table 13-1.

The development proceeds by first introducing the phases of the contention resolution (or arbitration) protocol phase by phase. Then the model is transformed to be considered cycle-accurate. The next design steps consider the introduction of local state, of inputs and outputs, and of an implementable bus model. The remaining steps are auxiliary to ease proof and to prepare recomposition. The development process is summed up on the figure 13-1. Introduction of protocol phases is done in refinements 0 to 5. The synchronization is added into the model by the refinement 6. Localization is done in refinements 7 to 14. Refinements 15 to 17 introduce inputs and outputs explicitly.

Table 13-1. Proof effort

Refinement	PO	Interactive proofs	Automatic proofs	Size of proof script	Calls to the automatic prover	Average size
0	2	0	2			
1st	5	2	3	9	1	4,5
2nd	19	5	14	31	5	6,2
3rd	5	5	0	28	4	5,6
4th	18	13	5	58	15	4,5
5th	14	5	9	32	3	6,4
6th	504	211	293	2660	744	12,6
7th	64	50	14	354	78	7,1
8th	36	28	8	150	46	5,4
9th	52	30	22	225	51	7,5
10th	380	176	204	1646	441	9,4
11th	31	23	8	109	15	4,7
12th	24	20	4	96	19	4,8
13th	75	33	42	673	172	20,4
14th	55	30	25	324	79	10,8
15th	46	34	12	167	32	4,9
16th	72	46	26	233	40	5
17th	87	59	28	445	132	7,5
Total	1489	51,7%	48,3%			9,4

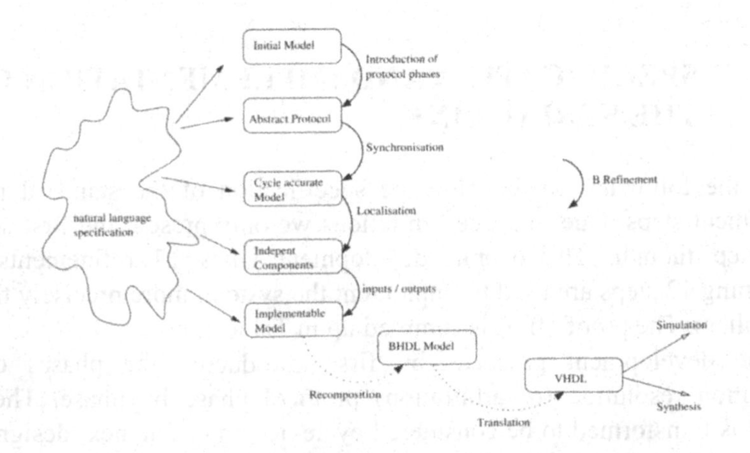

Figure 13-1. Hardware development process in Event-B

4.1 Abstract system

This machine models the bus as it would be seen by an observer, who is only able to see the bus. The system consists of a bus and several components represented by their controllers. The abstract machine models the most important property of the system. We just consider whether the bus is free or busy.

4.1.1 The state

SETS
ND
VARIABLES
wr
INVARIANT
$wr \subseteq ND \wedge$
$\forall\ (xx,yy).(xx \in wr \wedge yy \in wr \Rightarrow xx{=}yy)$

Figure 13-2. State of the abstract machine

The set *ND* models the components that are connected to the bus. The state of the bus is modeled by variable *wr*. If non-empty, the value of *wr* corresponds to the component that is writing to the bus. The set *wr* is empty if the bus is free.

In this first abstract system we specify only that the bus may be used by at most one component at any moment. There are two events that express taking possession of the bus (event *take*) and giving up ownership of the bus (event *free*).

The invariant ensures that *wr* is either empty or a singleton (the invariant is equivalent to $wr{=}\varnothing \vee \exists xx.(xx \in ND \wedge wr{=}\{xx\})$), **to ensure that at most one component writes to the bus at any time.**

4.1.2 Events

INITIALISATION
$wr := \emptyset$

Figure 13-3. Initialisation of the abstract machine

The initialization sets *wr* empty: when the system resets, no component is writing the bus.

```
take =
  ANY nd WHERE
    wr = ∅ ∧
    nd ∈ ND
  THEN
    wr := {nd}
  END
```

Figure 13-4. Event *take*

The event *take* may occur when the bus is free. On its occurrence a component *nd* takes possession of the bus.

For now, it is just specified that a component is chosen. The way this choice is done is the object of the development and will be made more precise in following refinements.

```
free =
  SELECT wr ≠ ∅ THEN
    wr:= ∅
  END
```

Figure 13-5. Event *free*

Event *free* may occur when the bus is busy. It models the passage of the idle-time after a bus-access. After its occurrence, the idle phase is completed and the bus is hence considered free.

```
nothing =
  ANY nd WHERE
    nd ∈ ND ∧ nd ∉ wr
  THEN
    skip
  END
```

Figure 13-6. Event nothing

The event *nothing* models the behavior of components that are doing nothing. The substitution *skip* is the substitution which changes nothing. This event cannot occur for the component which is using the bus. So, the only event that can occur for this component is *free*. This ensures that the bus will be always freed.

4.2 Specification of Protocol for Contention Resolution

We introduce the five phases of the contention resolution protocol (see figure below) specified in the standard in five refinement steps. This happens still at an abstract level.

- waiting phase: components that want bus-access are waiting,
- elimination phase: eliminates components that are still waiting when the waiting phase has finished,
- message phase: C sends its message,

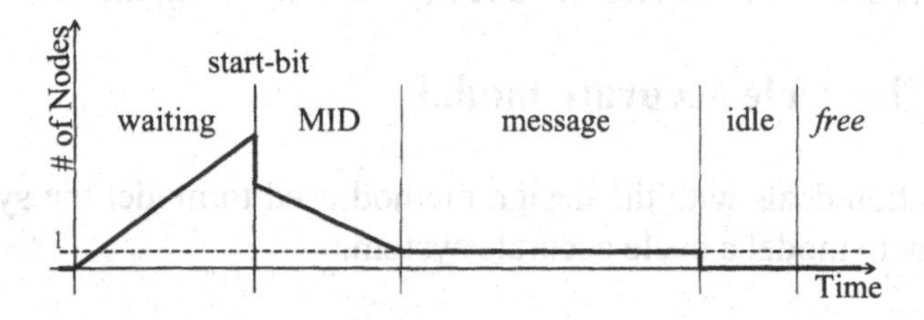

Figure 13-7. The five phases of the protocol

- Idle phase, following message phase: components are waiting for an idle line.

State of the system is composed of four variables:

1. the variable *wr* that represents the abstract bus
2. the partial function *AT*, it is a counter that exists only for component into the waiting phase,
3. the set *CP* is the set of component into the competition for accessing the bus, it decreases during competition,
4. an abstract variables *XR* that represents nodes that have been eliminated from the competition

All these variables are initialized to the empty set: when the system starts, there is no node using the bus, waiting or in the competition.

```
rise =
  ANY nd, tt WHERE
    CP = ∅ ∧ wr = ∅ ∧
    nd ∈ ND ∧
    nd ∉ dom(AT) ∧ tt ∈ ℕ1 ∧
    nd ∉ XR
  THEN
    AT := AT ⩤ {nd ↦ tt}
  END

rise_count =
  ANY nd WHERE
    CP = ∅ ∧ wr = ∅ ∧
    nd ∈ dom(AT) ∧
    AT(nd) ≠ 0
  THEN
    AT := AT ⩤ {nd ↦ AT(nd)-1}
  END
```

Figure 13-8. Events rise and rise_count

The B system obtained at this stage is composed of eight events. The figure 13-8 presents the two events *rise* and *rise_count* that model the waiting phase: *rise* adds a node to the set of waiting events and *rise_count* decreases the waiting counter (*AT*) of nodes that are waiting.

The five phases are modeled on an abstract level. In a final step, each component must see only its internal state and its input. This is why our goal is that the guard and the substitution of each event use only variables that the component may access. This is done by distributing the protocol. Components can not communicate directly but only using the bus.

4.3 The cycle accurate model

This section deals with the design method used to model the system. We explain how to model a cycle accurate system.

$$CP = \emptyset \wedge wr = \emptyset \wedge$$

- - - - - - - - - - - - - -

$$nd \in \mathbf{dom}(AT) \wedge$$
$$AT(nd) \neq 0$$

Figure 13-9. Guards of rise_count decomposed in two parts

Of course, the model has to take into account the concurrent execution of components. The concurrency is modeled using the above *ANY nd WHERE* form of events. The guard has two parts: the first part describes the condition under which the event may occur, and the second part defines which components are concerned by the event. For example, the guard of rise count, in the last refinement may be decomposed in two parts (cf. figure 13-9).

The first part of the guard expresses that the event may occur when there is no component in the competition phase ($CP = \emptyset$) and the bus is free ($wr = \emptyset$). The second part expresses that only components which are still waiting ($nd \in dom(AT) \wedge AT(nd) \neq 0$) are concerned. If the event occurs, one of these components is chosen non-deterministically.

We choose a cycle accurate model. At each cycle, an event has to occur for each component. If we want a deterministic model, it is required that just one event may occur for each component, at each cycle.

At each cycle, a component has to do two things: it must write the bus and read the bus. Because of the concurrent execution of components, and the fact the bus value is the result of all bus accesses, we have to separate a writing phase and a reading phase. We must make sure that all components have written to the bus, before reading it, so that the bus value is valid when reading commences.

To model this behavior, a two-phase cycle model is used. During the first, each component writes its output to the bus. The bus must not be read during this phase, because the value on the bus is not stable (not all components have written their output yet). During the second phase, every component may read the bus.

$$
\begin{aligned}
&VER \subseteq ND \wedge \\
&WRI \subseteq (ND\text{-}VER) \wedge \\
&VER \subseteq (ND\text{-}WRI)
\end{aligned}
$$

Figure 13-10. Two extra variables for synchronization

To specify this two-phase cycle model, we use two extra variables (*WRI* and *VER*), an extra event (*env*) and we have to add *synchronization* into each event's guard. This is done using tokens.

```
env =
  SELECT
    VER = ∅ ∧ WRI = ∅
  THEN
    S ||
    WRI := ND
  END
```

Figure 13-11. The event env

The variable *WRI* contains all components that are in the writing phase and *VER* the ones that have written to the bus and are ready to commence the reading phase. When a component writes the bus, it removes its token from *WRI* and enters it in *VER*. While *WRI* is not empty the system is in the writing phase. As soon as *WRI* is empty the reading phase commences.

```
Event_write =
  ANY nd WHERE
    nd ∈ WRI
    GUARD(nd)
  THEN
    S ||
    VER := VER ∪ {nd} ||
    WRI := WRI - {nd}
  END
```

Figure 13-12. Template for events of the writing phase

```
Event_read =
  ANY nd WHERE
    nd ∈ VER ∧ WRI = ∅ ∧
    GUARD(nd)
  THEN
    S ||
    VER := VER - {nd}
  END
```

Figure 13-13. Template for events of the reading phase

The event *env* is added. Event *env* does not express the behavior of a component. It models the environment of the circuit. We use this event to model the transition from one cycle to the next. The event *env* "initializes" the cycle. At the beginning, all components are in the writing phase($WRI := ND$ in env) .The behavior of each component must be modeled. This means that there must exist an event which may occur for each component. When an event occurs for a component *nd*, *nd* is removed from the writing phase ($WRI := WRI - \{nd\}$) and put in the set representing the components that are ready for the reading phase ($VER := VER \cup \{nd\}$).

When all components have completed the writing phase ($WRI = \varnothing$), the reading can start. As in the writing phase, an event must occur for each component and each event removes the component from the reading phase ($VER := VER - \{nd\}$).

When the reading phase is completed ($VER = \varnothing \wedge WRI = \varnothing$), the event *env* may occur again and a new cycle commences. Thus, events can be of three kinds: the event *env*, events of the *writing phase* and events of the *reading phase*.

4.4 Bit-Level definitions

For implementation purposes the abstract data representations of early stages must be made more concrete. At this level we introduce a type to represent bits, and implementation types of numbers as bit-arrays.

DEFINITIONS

BIT == {0,1}

CONSTANTS

MESSAGES, CHAR, MID, MAX_BIT_CHAR, ...

PROPERTIES

$MAX\_BIT\_CHAR \in N_1 \wedge$

$CHAR = 0..MAX\_BIT\_CHAR \rightarrow BIT \wedge$

$\forall cc.(cc \in CHAR \Rightarrow cc(0) = 0 \wedge cc(MAX\_BIT\_CHAR) = 1) \wedge ....$

The bus is an analog part of the system, its behavior consist in carrying 0 if one of the components is writing 0 and carrying 1 in other cases, this means that the bus carries the minimum of written values. We model this by using the pattern *bus* := *min*(*bus*, *output*) to write the bus. Latter, this is refined by modeling the bus itself.

4.5 Concurrency

```
ANY nd WHERE
  P(nd, v)
THEN
  S(nd, v)
END
```

Figure 13-14. The *ANY nd WHERE* form of an event

The behavior of components is modeled by events. Each event corresponds to the behavior of a set of components in a particular situation.

Events are of the *ANY nd WHERE* form. The predicate *P*(*nd*, *v*) expresses the condition under which the event may occur and also constraints on *nd*. Constraints on *nd* define a set of components (cf. section 4.3). When the event occurs, one component *nd* is chosen non-deterministically (wrt. *P*). When more than one event may occur, the choice is also non deterministic.

Furthermore, the set of events must describe the behavior of all components, in all cases. At each cycle, each component must do something. So, the collection of events is like if there was one event for each component. The independence of components and constraints on local state ensures that when an event occurs for a component, it does not modify the state of other components. So, in whatever order the events may occur, the result on the state of the components is the same. This is a model of the concurrency. We must also ensure that the behavior of the system is correctly synchronized, i.e. all components are in the same cycle at the same instant.

4.6 Local state

Each component has to be separately modeled. Each component must have its own state such that the components are independent. Because the number of components is not fixed, we model states of components by arrays, represented in B by total functions. Each index in the array corresponds to a component. For example, to model a state variable *vstate* for all components, we use the invariant $vstate \in ND \rightarrow V$. Where *ND* is the set of connected components and *V* the set of acceptable values of *vstate*.

Each component is independent, then for each event, apart from the event *env* which does not express the behavior of a component, the guard and the substitution must depend only of the state and inputs of the component that is treated. Because all events are of the *ANY nd WHERE* form, the guard of the substitution depends only on *nd* and on the bus state (input of components).

In this study we do not descend to the bit level to model local states. Local states are modeled using sets. To be considered as modeling local state, these sets have to be handled with precaution. In an event of the *ANY nd WHERE*, such a set can be handled only with a substitution of the form $Set := Set-\{nd\}$ or $Set := Set \cup \{nd\}$. This corresponds to model the set, at a bit level, by a local variable of type bit. The substitution $Set := Set \cup \{nd\}$ corresponds to setting the variable to *true* and $Set := Set-\{nd\}$ to setting the variable to *false*. In the same way, the only acceptable way of testing such a set is $nd \in Set$ or $nd \notin Set$. At a bit level, this corresponds to test whether the bit variable is *true* or *false*. In particular, tests of the form $Set = \varnothing$ are not allowed.

During the development, a refinement does this transformation to obtain a synthesizable model. Each set *Set* is refined by a boolean value *BoolSet*, a gluing invariant specifies that $BoolSet(nd)=true \Leftrightarrow nd \in Set$.

These constraints apply for all events, apart from *env*. However, it is impossible to do the synchronization using only local guards and substitution. So, there is an exception about extra variables *VER* and *WRI*. These variables are removed (the synchronization is implemented) when the model in expressed in BHDL.

4.7 Input and Output

During the development inputs are modeled abstractly by non-deterministic choice over all possible values. To be able to translate to VHDL, the inputs must be modeled explicitly though.

In the B model, inputs are simply variables. However they are never assigned a particular value. An input can be modified only at two places: in the initialization (because all variables must have an initialization in a B model) and in the event *env*. Moreover we have said also that the event *env* models the transition from one cycle to another. The way the event *env* handles input variables is a model of the environment, it can be used as a specification of the interface of the system.

4.8 Recomposition and translation

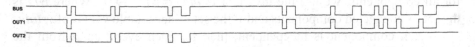

Figure 13-15. Simulation of the VHDL code

At the end of the refinement process, the eventB model is recomposed (cf. chapter 16). Events of the model are composed together to obtain the code of a controller in one block. The model obtained by recomposition is expressed in BHDL©. From the BHDL model, we have used the translator to obtain the VHDL description. Figure 13-15 give an example of simulation.

By simulation of the original design we found a low-level fault: the circuit contained a glitch. The proposed method does not **consider this type of low-level fault. Thus it has to be analyzed by other means like simulation. It would not be difficult however to implement a tool that warns about a possible glitch.**

4.9 Final architecture

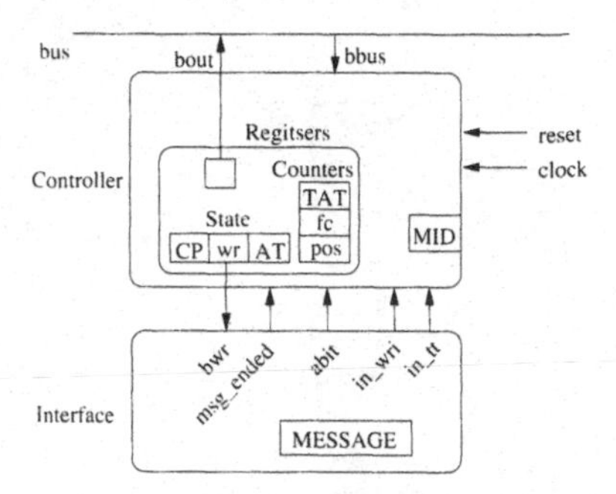

Figure 13-16. Schema of the controller

The controller we have developed by formal refinement in the B method is schematized in the figure below. The BHDL description has been translated into VHDL and synthesized.

```
entity ControlRev is
port (
clock : in BIT;
reset : in BIT;
bbus : in BOOL;
in_tt : in UINT4;
in_wri : in BOOL;
MID : in CHAR;
abit : in BOOL;
msgended : in BOOL;
bout : out BOOL;
bwr : out BOOL
);
end;
```

5. USING UML-B

The B model was difficult to comprehend because it is not an entirely natural representation of real entities. The model was visualized by manual translation into UML-B.

Our first UML-B model tried to replicate the B model giving the following state chart (representing the state of all nodes). The state chart is a redundant alternative representation since the 'real' state is captured in the variables WA etc. The diagram provided a mechanism for a group discussion of the model. In particular we discussed the elimination phase, which is an artifact of the modeling of concurrency and not a real phase for a particular node.

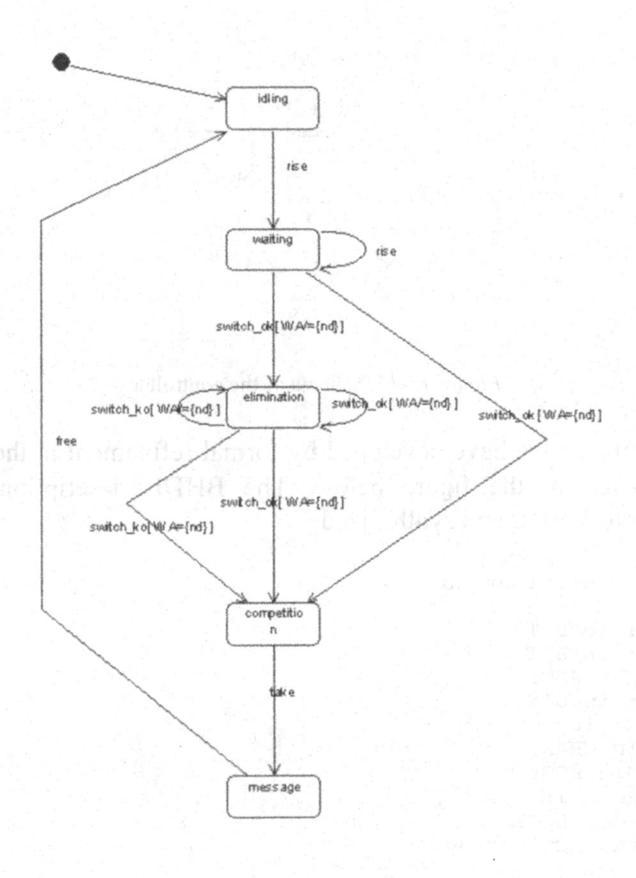

Figure 13-17. colin1

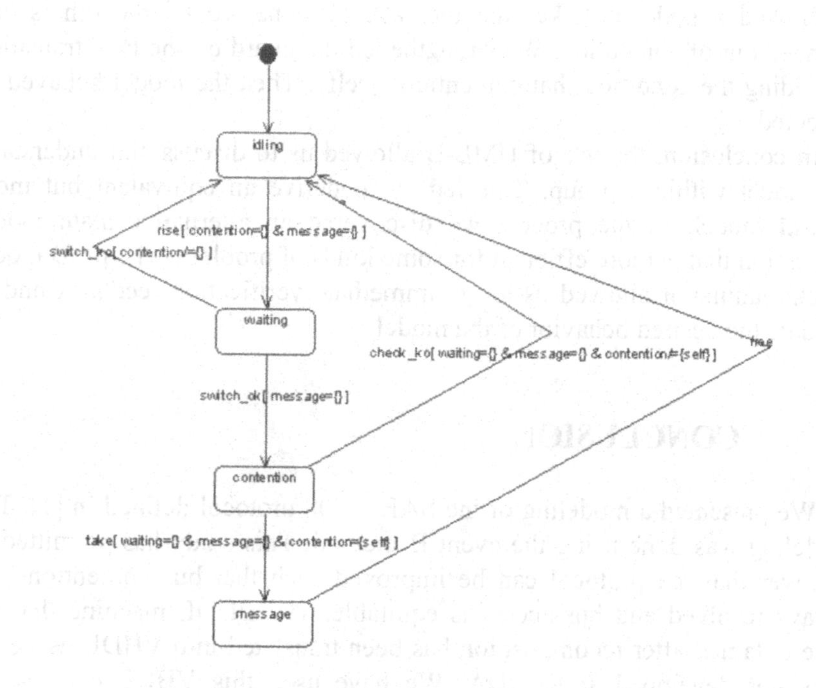

Figure 13-18. colin2

We decided to re-model the system using the node class to base the state chart. The class diagram is trivial with a single class, NODES. This gave a simpler state model. Each node instance owns an instance of the state model. The elimination phase is implicit in the distribution of the state value of nodes. However, with the semantics we had at the time, the state model represents a single value that can take on the value of each state (i.e. the states are an enumerated set forming the type of the state variable). In this example we found that this lead to a cumbersome expression of guards involving inversion of functions resulting in difficult proofs. e.g. SS~[{waiting}]={} & SS~[{message}]={} & SS~[{contention}]={self}.

We changed the state model semantics and U2B translation so that each state represents a variable that is the set of node instances currently in that state. Initially idling=all nodes and all other states are empty. When a transition event occurs, the node is removed from the starting state set and added to the target state set. This resulted in a simpler expression of guards and removed the proof difficulties.

The ProB [4] model checker was used to verify the model. ProB detected a deadlock because we had omitted the transition 'free' in the state model. After correcting this, the model was verified correctly. We then ran the ProB animator. We discovered that the model did not behave as expected because it allowed a node to 'take' the message state before all the others had dropped out of contention. We strengthened the guard on the take transition by adding the condition that contention={self}. Then the model behaved as expected.

In conclusion, the use of UML-B allowed us to discuss and understand the model within a group. This led us to derive an equivalent but more natural model. In the process we discovered an alternative state model translation that is more efficient for some kinds of problem. The proB model checker/animator allowed us to get immediate verification feedback and to validate the desired behavior of the model.

6. CONCLUSION

We presented a modeling of the SAE J1708 protocol defined in [1]. The modeling was done using the event B method. This study has permitted to discover that the protocol can be improved such that bus contentions are always resolved and bus-access is equitable. The BHDL machine that we have obtained after recomposition has been translated into VHDL using the translator developed at KeesDA. We have used this VHDL code to do simulation. In our first recomposition attempt this simulation shown glitches on output signals of the circuit. This was corrected by doing another recomposition following design guidelines given by Volvo. The design was then simulated and successfully synthesized.

ACKNOWLEGMENT

We are grateful to Ola Lundkvist and Eilert Johansson at Volvo for their king help. We are also grateful to Marcin Zys at Evatronix for running the synthesis of our code.

REFERENCES

[1] SAE International. SAE J1708 revised OCT93, serial data communication between microcomputer systems in heavy-duty vehicle, http://www.sae.org

[2] Jean-Raymond Abrial. Event Driven Electronic Circuit Construction, August 2001

[3] J.-R. Abrial and L. Mussat. Introducing dynamic constraints in B. In B'98: Recent Advances in the development and Use of the B Method, volume 1393 of LNCS, 1998

[4] Michael Leuschel and Michael Butler. ProB: A Model Checker for B. In FME 2003: Formal Methods, volume 2805 of LNCS, 2003

Chapter 14

THE ECHO CANCELLATION UNIT CASE STUDY

Alexander Krupp, Wolfgang Mueller
Paderborn University, Paderborn, Germany

Ian Oliver
Nokia Research Centre, Helsinki, Finland

Abstract This chapter describes a case study, which was performed in collaboration of Nokia Research and Paderborn University in the context of the methodology for refinement automation of finite state machines as presented in Chapter 10. We elaborate on a combined verification and refinement of an echo cancellation unit of a mobile phone with the Atelier-B toolkit and the RAVEN model checker. The experimental results show that the presented methodology requires almost no user interaction and gives acceptable runtimes with Atelier-B.

Keywords: formal refinement, model checking, finite state machines, real-time systems, echo cancellation unit

1. Introduction

One of the main drawbacks of theorem proving is still the frequently required manual interaction by highly skilled engineers with deep understanding of the model under refinement. To overcome this, Chapter 10 introduces a coding style and a methodology for the specification, refinement, and model checking of finite state machines with the Atelier-B theorem prover [1, 3] and the RAVEN model checker [2], respectively. Along the lines of classical HW/SW co-design, we investigated the re-

J. Mermet (ed.), UML-B Specification for Proven Embedded System Design, 227–240.
© 2004 *Kluwer Academic Publishers. Printed in the Netherlands.*

finement and verification from a cycle-accurate via time-accurate finite state machines to C code generation. Starting from state diagram based RAVEN models as given in Figure 1.1, we generate code for the Atelier-B theorem prover and enforce a coding style, which is oriented towards the basic patterns of RAVEN's I/O Interval Structures, i.e., synchronously communicating finite state machines (FSMs).

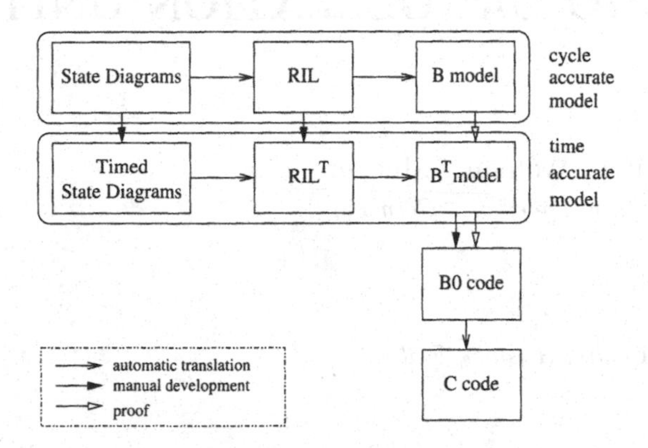

Figure 1.1. Combined Formal Refinement and Model Checking of FSMs

To achieve higher degrees of automation to significantly reduce manual interaction and decrease the theorem prover runtime, we have introduced a formal verification with B through refinement levels denoted as S-, SB-, and SBT-level. Those levels in succession add structural, behavioural, and timing properties to the system model. In that context, we have defined specific B patterns to separate the specification into easily provable parts for refinement automation, which basically refer to fully automatic refinement and low runtimes for the Atelier-B theorem prover. The following chapter applies the presented methodology for an industrial case study, namely, the echo cancellation unit (ECU) of a mobile phone. The case study gives first promising results for refinement automation as well as verification runtime performance.

The remainder of this chapter is organised in three sections. The next section introduces the ECU example. Thereafter, we elaborate on model checking and formal refinement in separate sections.

2. The Echo Cancellation Unit

Today, any mobile phone has an Echo Cancellation Unit (ECU) to digitally filter input and output streams to suppress crosstalk between speaker and microphone. The input audio streams come from the external speaker and microphone and go through a decoder from the GSM network. The ECU receives decoded audio from the GSM network and the speaker, simultaneously filters, and writes audio packets to the GSM net. The output stream goes from the network to the external speaker after decoding. The collaboration diagram in Figure 1.2 gives an overview of the ECU architecture and its functional components. As depicted there, the ECU resembles a producer-consumer synchronisation problem with two producers, two consumers, three buffers, and a filter, which can be considered as a producer and a consumer again.

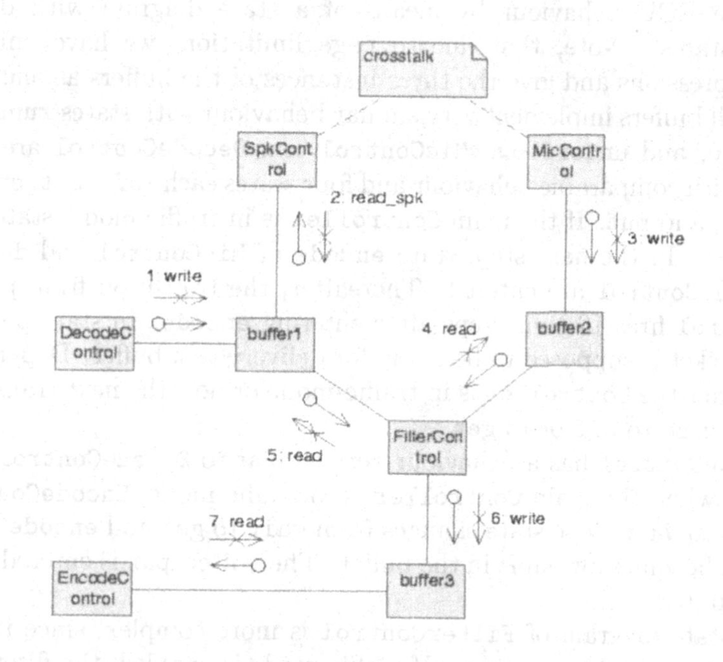

Figure 1.2. Collaboration Diagram for the Echo Cancellation Unit

An ECU decoding process `DecodeControl` decodes audio data packets received from the GSM network and inserts the packets to `buffer1`. The buffer duplicates audio packets to the speaker through `SpkControl` and to the digital `FilterControl`. `MicControl` receives audio from the microphone sending them to `buffer2`. The digital filter `FilterControl` receives packets from `buffer1and` `buffer2` and forwards the output to `buffer3`. The encoding `EncodeControl` takes the input from `buffer3` and releases the output to the GSM network.

For the correct function of an ECU, buffer over- and underflows are of special interest. For over- and underflow, we can identify the following conditions:

- An object is inserted into the buffer only if the buffer is not full. Otherwise, there is a buffer overflow.
- The consumer becomes active only if there is at least one object in the buffer. Otherwise, there is a buffer underflow.
- If read and write operations are performed at the same time, an empty buffer creates an underflow and a full buffer creates an overflow.

Complementary to the previous collaboration diagram, Figure 1.3 gives the ECU behaviour by means of a state diagram with different substates. Note, that due to page limitation, we have omitted some expressions and give the three instances of the buffers as one substate. All buffers implement very similar behaviour with states `running`, `overflow`, and `underflow`. `MicControl` and `DecodeControl` are producers with comparable behaviour and four states each: `off`, `get`, `encode /decode`, and `put`. If the main `Controller` is in traffic mode, state `get` is reached. In the next step, state `encode` of `MicControl` and `decode` of `DecodeControl` are entered. Thereafter, the transition from `put` of `MicControl` fires 10 time steps after entering `encode`. In state `put`, an audio packet is supposed to be ready for delivery to a buffer. Depending on whether the `Controller` is in traffic-mode or not, the next transition will either go to `off` or to `get`.

`EncodeControl` has a behaviour very similar to `DecodeControl`. After `off`, when the main `Controller` is in traffic mode, `EncodeControl` switches to `wait`. The state changes from `wait` to `get` and `encode` when a packet becomes available in the buffer. Thereafter, `put` is entered after 9 time units.

The state diagram of `FilterControl` is more complex, since it acts as a producer and a consumer. If traffic mode is enabled, the filter first executes `startup` and goes to `wait` just 1 time step after that. When a packet becomes available in `buffer1` and `buffer2`, the filter switches to `get`, `decode`, and proceeds to `put` just after 10 time units, where a packet is inserted into `buffer3`. If traffic mode is still enabled, the filter returns to `wait`, otherwise to `waitflush` and to `off`.

We take the previous collaboration and state diagram as starting points for our formal verification. In the analysis of the ECU case study, we are mainly interested in

- the dimensioning of buffers to avoid under- and overflows,
- the correct synchronisation of modules when processing the audio packets,

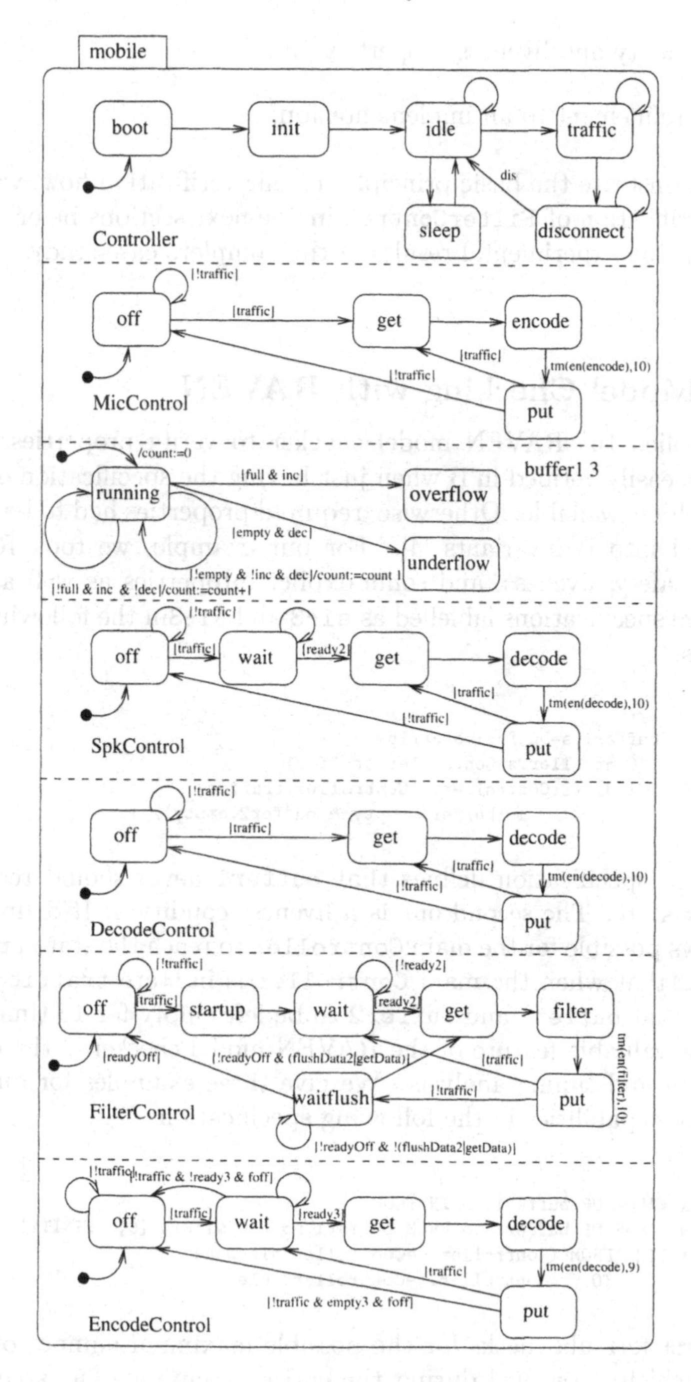

Figure 1.3. Timed State Diagram for ECU

- the safety and liveness properties, and
- the refinement to an implementation.

To demonstrate the basic principles of our verification flow, we focus on the verification of `FilterControl` in the next sections before finally presenting the experimental results of the complete case study.

3. Model Checking with RAVEN

We applied the RAVEN model checker to verify properties, which cannot be easily verified in B when just having the specification of finite state machine available. Otherwise, required properties had to be deeply introduced into B invariants [4]. For our example, we took RAVEN to verify safety, liveness, and some explicit properties as well as some quantative specifications lablelled as `s1-3` and `a1-3`in the following code fragments.

```
SPEC
  s1 := AG !(buffer1.s=buffer1.overflow )
  s2 := AG EF (Controller.s=Controller.traffic)
  s3 := AG( !EG[13] ( (Controller.s=Controller.traffic)
                    & (buffer1.empty & buffer2.empty)) )
```

The first specification defines that `buffer1` never should reach the `overflow` state. The second one is a liveness condition. It defines that it is always possible for the main `Controller` to reach the state `traffic`. `s3` defines that when the main `Controller` is in state `traffic`, we do not want that `buffer1` and `buffer2` to be left empty for 13 time steps.

A most valuable feature of the RAVEN model checker is the quantitative data and timing analysis. We give three examples for our ECU case study capabilities in the following specifications.

```
ANALYSE
  a1 := MAX VALUE OF buffer1.no IN TRUE
  a2 := MIN VALUE OF buffer1.no FROM buffer1.no > 0 WITHIN [0,INFINITY]
  a3 := MAX TIME FROM (Controller.s=Controller.disconnect)
                 TO   (Controller.s=Controller.idle)
```

The first formula checks for the possible maximum number of buffer objects, which is reached during the entire execution. The second one returns the minimum number of objects in `buffer1` after one object has been consumed, i.e., it has increased 0 objects. `a3` gives the maximum execution time, from `disconnect` to `idle` for the `Controller`.

Considering the previous, very simple specification, a verification run with RAVEN on the complete case study executes in 0.47 seconds under Linux on an Intel P4, 2.2GHz with 1GB main memory, for instance.

4. Formal Refinement with Atelier-B

We now outline structural, behavioural, and timed behavioural specifications for our case study before we give the concrete numbers of our experiments.

4.1 Different Levels of Refinement

Chapter 10 introduces a methodology for the refinement of finite state machines to timed finite state machines as given in Figure 1.4. Based on the three refinement levels, we presented a translation, which transforms finite state machines into B language code to conduct efficient proofs with the Atelier-B theorem prover.

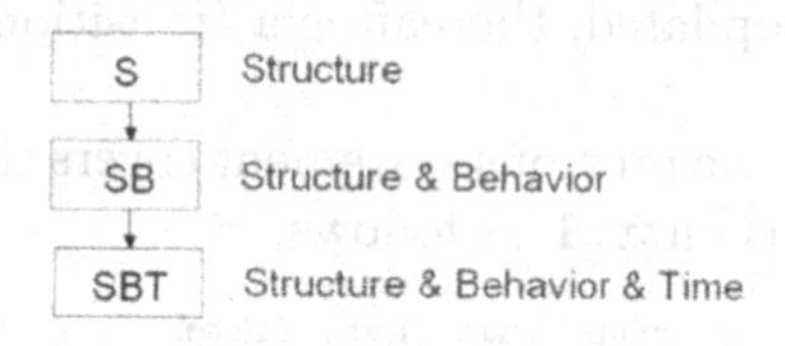

Figure 1.4. Different Levels of Refinement

The following paragraphs introduce the different S-, SB-, and SBT-level specifications for the ECU case study.

S-Level. Our S-level specification starts from the `MAIN` B machine, which recursively includes other machines. It mainly covers the specification of the underlying model of computation (MoC) for the embedded state machines. In our case, the MoC defines the execution model of the RAVEN model checker. However, we can include any other MoC here.

For each of the state machines modules, we have to define the communication structure for value propagation from outputs to inputs of the connected modules or state machines, respectively. For each state machine, the definition of state transitions is mainly kept undefined, i.e., a non-deterministic selection over all possible states is defined.

The following B specification sketches the S-level definition of the `FilterControl` state machine with its operations `propagateValues` for value propagation and `doTransition` for state transitions.

```
MACHINE FilterControl_S
...
OPERATIONS
  propagateValues( ... ) =
  PRE
  ...
  THEN
  ...|| valuesUpdated := TRUE
  END;

  doTransition =
  PRE
    valuesUpdated = TRUE
  THEN
    SIGNAL_s :: SIGNAL_T_s || valuesUpdated := FALSE
  END;
```

doTransition just defines a non-deterministic selection of the value of the state variable SIGNAL_s from all of its possible values given in the type definition SIGNAL_T_s. Note here, that the variable identifier reflects its correspondence to signal s of the RAVEN model. In the above code, variable valuesUpdated additionally guarantees a mutual exclusive execution between both operations. That means, values are first propagated and updated, thereafter a transition fires, then, values are updated again etc.

The type definition the previous statement covers the set of all possible state values of FilterControl as follows.

```
SIGNAL_T_s = {off, startup, wait, get, filter, put, waitflush}
```

SB-Level. The SB-level specification is a refinement of the doTransition operation. It refines the operation by adding state transitions details. That means, to replace the general non-deterministic state selection by deterministic elements or by limited sort of non-determinism. The following example shows the SB-level definition of two state transitions; from off to startup and from filter to put.

```
REFINEMENT FilterControl_SB
REFINES FilterControl_S
...
OPERATIONS
  doTransition =
  BEGIN
    IF ( SIGNAL_s = off ) THEN
      SELECT ( INPUT_traffic = TRUE ) THEN
        SIGNAL_s := startup
      ...
    END
    ELSIF ( SIGNAL_s = filter ) THEN
      SELECT ( TRUE = TRUE ) THEN
        SIGNAL_s := put
      WHEN ( TRUE = TRUE ) THEN
```

```
        SIGNAL_s := SIGNAL_s
      END
    ...
  END
```

The specification starts with a definition of a transition from `off` to `startup` when traffic is indicated by the controller. The traffic variable is an input variable from another module as indicated the variable's prefix. From `filter`, a non-deterministic transition to `put` or `filter` is specified. The non-deterministic transition is meant to be replaced in a later refinement by a timed transition.

SBT-Level. SBT-level further refines the `doTransition` operation. Modifications are the replacement of transitions by timed deterministic transitions and the inclusion of a `Timer`. The following B code demonstrates how the transition from `filter` to `put` of previous B code is refined to a timed transition.

```
REFINEMENT FilterControl_SBT
REFINES FilterControl_SB
INCLUDES t1.Timer(10)
...
OPERATIONS
  doTransition =
  BEGIN
    ...
    ELSIF ( SIGNAL_s = filter ) THEN
      IF (t1.elapsed = TRUE) THEN
        SIGNAL_s := put || t1.doReset
      ELSE
        t1.doAdvance
      END
    ...
  END
END
```

We first include a timer `t1` for modelling the 10 time units delay. Thereafter, the refined transition is encapsulated with statements advancing and resetting `t1`. A guard first checks for the elapsed timer. When not elapsed, `t1` is advanced in each call of `doTransition`, i.e., in each time step. When the timer elapses, the transition is executed and the timer is reset simultaneously.

For a comprehensive representation of the B code and to achieve a better readability, we may also use the following macro, which hides implementation details of the encapsulation. Here, the assignment of the target state is encapsulated by a `DELAY ... THEN ... END` statement.

```
...
ELSIF ( SIGNAL_s = filter ) THEN
  DELAY 10 THEN SIGNAL_s := put END
```

4.2 Experimental Results

Successful proof of the composed system in B very much depends on the adequate separation of the individual modules into B machines. Here, we give numbers of Atelier-B, which were achieved with the separation into the previously introduced refinement levels. Refinement with Atelier-B basically proves the range of variables via the invariant of each machine and the refinement of transitions via B refinement rules for substitutions.

Component	Obvious POs	Non Obvious POs	Interactive	Unproven
S-Level:				
MAIN_S	8	0	0	0
GLOBAL_CONST	1	0	0	0
Controller_S	35	0	0	0
DecodeControl_S	29	0	0	0
EncodeControl_S	38	0	0	0
FilterControl_S	77	0	0	0
MicControl_S	29	0	0	0
SpkControl_S	32	0	0	0
Timer	31	7	1	0
buffer1_S	34	1	0	0
buffer2_S	40	1	0	0
buffer3_S	34	1	0	0
SB-Level:				
Controller_SB	156	0	0	0
DecodeControl_SB	64	0	0	0
EncodeControl_SB	178	0	0	0
FilterControl_SB	481	0	0	0
MicControl_SB	64	0	0	0
SBT-Level:				
Controller_SBT	223	2	0	0
DecodeControl_SBT	137	0	0	0
EncodeControl_SBT	320	1	0	0
FilterControl_SBT	746	5	0	0
MicControl_SBT	137	0	0	0
TOTAL	2894	18	1	0

The above table gives a summary of the proof results with Atelier-B 3.6b. It gives the number of obvious and non obvious proof obligations (POs), i.e., generated proofs, for each B machine and shows that the complete proof was almost fully automatic. No unproven property was left and just one very simple manual interaction with the invocation of a short command was required: `pp(rp.0)`, which executes a simple invocation of the predicate prover, where the parameter defines that the prover shall ignore all hypotheses.

The complete proof was performed on an Intel P4, 2.2GHz, 1GB RAM. The proof of proof obligations in *Force 1* mode of Atelier-B took approx. 291 s in total. The execution time of the predicate prover interactive command was less than a second.

The previous investigations give numbers strictly following the approach from structural (S-level) to timed behavioural specification (SBT-level). Starting from that approach we checked for alternatives in the refinement of finite state machines. For outlining the alternatives, we consider a further refinement for a low power extension of the SBT-level `FilterControl` specification as given in Figure 1.5.

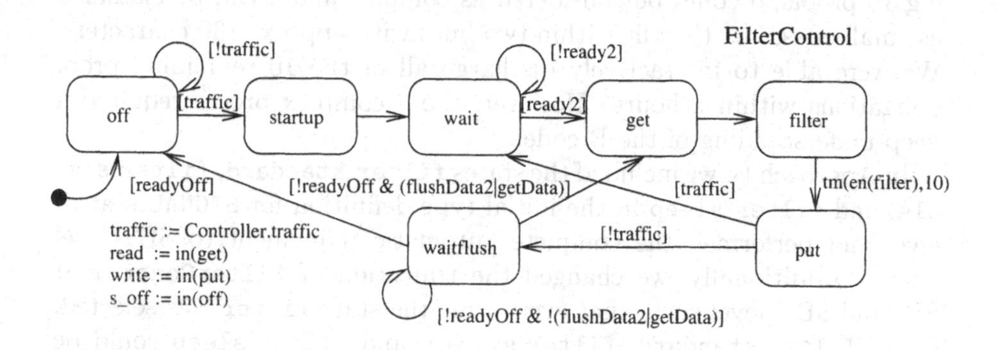

Figure 1.5. SBT-Level `FilterControl`

In the low power extension, a simpler filtering algorithm with less power consumption is selected when no significant data are transmitted, i.e., when the microphone volume is below a specific threshold. For that, we focus on the replacement of the `filter` state considering the output transition delay of 10 time units. The refined state machine switches between filtering algorithms by replacing state `filter` as it is shown in Figure 1.6. Thus, from `get`, either a standard filter or a simplified filter is selected. Since the processing of the simpler filter finishes earlier, the DSP is set to sleep mode for the remaining time within the previously specified time bounds of `10` time units.

We now investigate three different approaches for a low power refinement, which are denoted as Approach A, B, and C. In Approach A, we introduced the additional states `filter_standard`, `filter_simple` and `filter_sleep` only in the refined SBT-level machine. For that, the type of the state variable `SIGNAL_s` was replaced by a type, which covered the

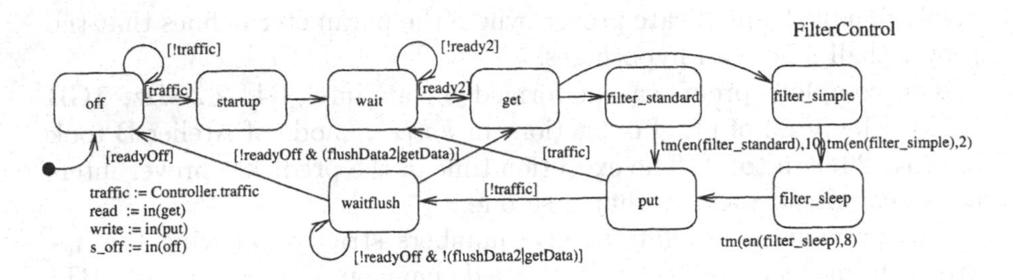

Figure 1.6. Refined SBT-Level **FilterControl**

extended state set and a linking invariant was added. After modifying the invariant, we achieved 14 proof obligations with large expressions. 4 were solved with a very simple interactive command. From the remaining 10 proofs, 6 could be considered as complex and 4 can be classified as small ones, i.e., they fit within two lines with approx. 130 characters. We were able to interactively discharge all of the 10 remaining proof obligations within 2 hours. However, the 6 complex proofs required a deep understanding of the B code.

In Approach B, we included the states **filter_standard**, **filter_simple**, and **filter_sleep** in the initial type definition for **SIGNAL_s** at S-level and performed the complete automatic refinement to SBT-level again. Additionally, we changed the transition of **FilterControl** at SB- and SBT-level such, that every time the state **filter** was selected, states **filter_standard**, **filter_simple**, and **filter_sleep** could be selected non-deterministically.

```
WHEN ( SIGNAL_s : {filter_standard,filter_simple,filter_sleep} ) THEN
   SELECT ( TRUE = TRUE ) THEN
      SIGNAL_s := put
   WHEN ( TRUE = TRUE ) THEN
      SIGNAL_s :: {filter_standard,filter_simple,filter_sleep}
   END
```

At SBT-level, when in one of the states **filter_standard**, **filter_simple**, or **filter_sleep**, we additionally increment the timer and allow a non-deterministic choice of the next state out of the three states. When the timer elapses, it is reset and state changes to **put** as follows.

```
WHEN ( SIGNAL_s : {filter_standard,filter_simple,filter_sleep} ) THEN
   IF (t1.elapsed = TRUE) THEN
      t1.doReset
   ELSE
      t1.doAdvance
   END                                    ||
   IF (t1.elapsed = TRUE) THEN
```

```
        SIGNAL_s := put
    ELSE
        SIGNAL_s :: {filter_standard,filter_simple,filter_sleep}
    END
```

When in state **get**, the next state is selected non-deterministically between **filter_standard**, and **filter_simple**. The complete proof of the refinement was almost automatically. That means, that just three POs had to be proven interactively by a very simple invocation of the rule based prover:

```
ff(1) & pr & pr & pr
```

that invokes the automatic prover with normal force (Force 1) and then executes the automatic prover three times. Note here, that Approach B is almost identical to Approach A except that we do not introduce a new state variable here.

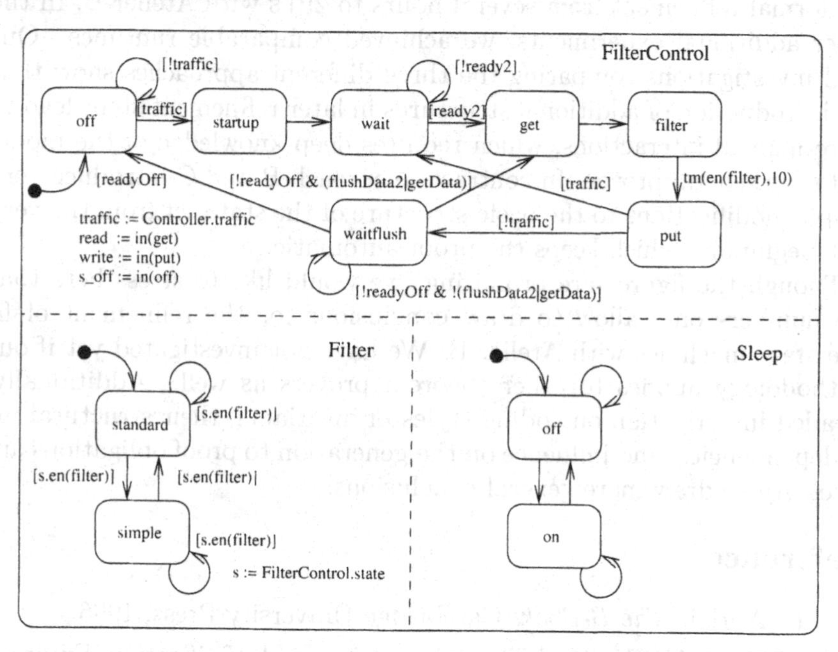

Figure 1.7. FilterControl refinement with Concurrent States

In Approach C, we added the same low power function by introducing two additional concurrent state machines as given in Figure 1.7. Here, we use a second timer `t2` for duration of the simple filter execution. `filter` is entered from state `get` with either the `simple` or `standard` filter selected. In the final approach, Atelier-B generated a lot more obvious proof obligations than in the previous approach. Nevertheless, Atelier-B has proven all proof obligations fully automatically again.

The following table gives an overview of the three additional experiments. It shows that Approach A leaves 10 prove obligations unproven. The other two led to fully automatic proofs with additional invocation of simple commands.

	Obvious POs	Non Obvious POs	Interactive	Unproven
Approach A	6609	612	15	10
Approach B	4572	169	4	0
Approach C	6255	93	1	0

In conclusion, we find our experimental results quite promising. With an adequate separation of specifications, we could decrease our time of the formal refinement from several hours to 291 s with Atelier-B. In the other additional experiments, we achieved comparable runtimes. Our final investigations comparing the three different approaches show that the introduction of additional structures in later refinement steps lead to more manual interactions, which requires deep knowledge of the model and the theorem prover. In contrast, Approach B and C introduce very simple modifications to the basic structure of the state set from the very first beginning, which keeps the proof automatic.

Though the figures are promising, we would like to stress here that the numbers only allow to draw conclusions for the refinement of finite state machines with Atelier-B. We have not investigated yet if our methodology applies to other theorem provers as well. Additionally, detailed investigation on coding styles of invariants, their structural interdependencies, and influence on the generation to proof obligations are necessary to draw more general conclusions.

References

[1] J.R. Abrial. *The B-Book.* Cambridge University Press, 1996.

[2] J. Ruf. RAVEN: Real-Time Analyzing and Verification Environment. *Journal on Universal Computer Science (J.UCS), Springer, Heidelberg*, February 2001.

[3] ClearSy. *Atelier-B Users Manual* - Version 3.6. Aix-en-Provence, France, 2003.

[4] J. Zandin. Non-Operational, Temporal Specification Using the B Method - A Cruise Controller Case Study. Master's Thesis, Dept. of Computing Science, Chalmers University of Technology and Gothenburg University, 1999.

Chapter 15

RESULTS OF THE MOBILE DESIGN SYSTEM EXPERIMENT

Ian Oliver, Klaus Kronlöf

Nokia Research Center
Itämerenkatu 11-13
Helsinki, Finland

ian.oliver@nokia.com, klaus.kronlof@nokia.com

1. INTRODUCTION

This case study involves the specification of a digital signal processor architecture as may be developed using MDA (Model Driven Architecture) techniques.

A key point to note here is that the use of formal specification and model based approaches to design are already well known and used widely within Nokia. We have a number of well defined approaches to developing in a MDA style and much experience (notably Octopus which is in the public domain) in the development of methodologies for system development.

The DSP case study was designed to investigate was how we could utilise the technologies developed within the PUSSEE project within our existing modelling frameworks and methods. As MDA is the latest trend in modelling frameworks then the ideas here we wished to integrate as well.

The case study then developed into a number of parts: DSP Architecture, Echo Cancellation Subsystem and the integration of those descriptions.

The DSP architecture starts off as a very high level design expressed using the UML-B profile. Properties such as internal consistency are checked and the model developed through refinement.

The Echo Cancellation Subsystem was developed using partially UML as the host notation but mainly utilising B and timing properties that are checked using

J. Mermet (ed.), UML-B Specification for Proven Embedded System Design, 241–259.
© 2004 *Kluwer Academic Publishers. Printed in the Netherlands.*

the RAVEN model checker. This particular part of this case study is decribed in more detail elsewhere in this book.

The model integration was designed to investigate the integration of the subsystem design with the higher level DSP architecture. This situation is common as a pure top-down design method is in practise almost impossible to execute. We also rely on existing components and subsystem designs - it is impractical, costly and risky to redesign systems every time. Reuse of components is critical and thus investigation of how one can utilise existing components within PUSSEE is done here.

1.1 DSP GSM Channel Management

Conceptually the architecture of a typical digital signal processor is not much more than the relationship between a controller module, protocols, logical channels and some hardware units for the reception or transmission of data.

In the context of the project the GSM Mouly and Pautet, 1992 channel data structure and its relationship to a single base station was modelled with the DSP transmission mode states.

The DSP transmission modes correspond to particular states that the DSP can be in at certain times. There are basically four modes or states: boot, initialise, idle and traffic. These correspond to the DSP boot sequence - obtaining details from the MCU regarding the CODECs available, what capabilties are required, memory locations etc.

The initialisation mode is related to the DSP obtaining from the 'network' the list of broadcasting channels, their properties and what data profiles are supported. If there are enough channels capable of supporting voice and/or data traffic available then the DSP can proceed to idle mode.

The idle mode corresponds to the situation where the DSP is fully configured and periodically requests data from the MCU or network. In this mode no active voice or data communication initiated by the user is made. During this mode it is possible that the DSP processor can switch into a power save mode.

The traffic mode corresponds to the situation where the DSP is fully configured and is actively involved in processing a voice or data call initiated by the user.

The development proceeded through three levels of refinement and mainly concentrated on issues regarding the modelling with UML-B. In the following sections we introduce the model and make a brief discussion on each.

1.2 Initial Models

In figures 1.1 and 1.2 we show an abstract representation of the DSP system, the channels and the cell in which the DSP has knowledge of. The state diagram (fig.1.2) shows in particular the four states (boot, init, idle and traffic) corresponding to the DSP modes.

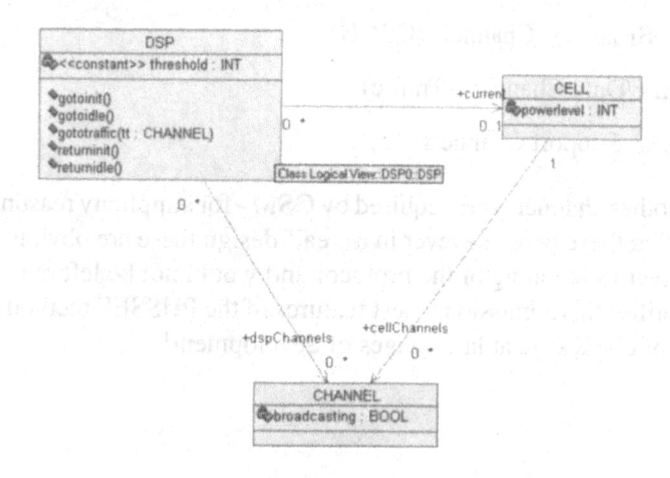

Figure 1.1. DSP Initial Model

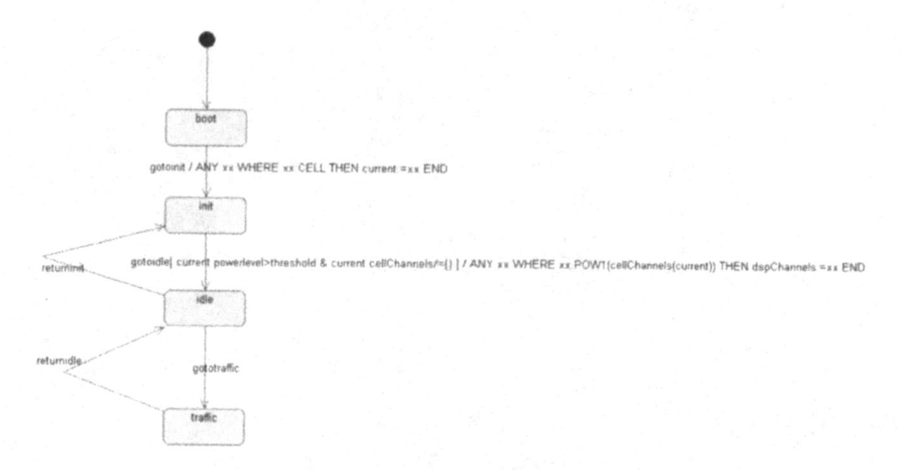

Figure 1.2. DSP Mode Diagram

In particular we introduce at this stage the basic operations in which the decision about the change of DSP state is made.

The notion of channel is abstract to say the least and in effect can model any channel based protocol and indeed in future design may even do so. Here however we already know that GSM is the target protocol.

Figure 1.3 shows the first refinement of the model described in figure 1.1. Of particular interest here is the subclassing of Channel into individual channel structures, primarily:

- Cell Broadcast Channels (CBCH)
- Traffic/Data Channels (Traffic)
- Access Support Channels

There are other channel types required by GSM - for simplicity reasons we have not modelled these here, however in a 'real' design these are obviously critical to the correct functioning of the protocol and would not be left out. We could however utilise this admission to test features of the PUSSEE method regarding additions of classes etc at later stages of development[1].

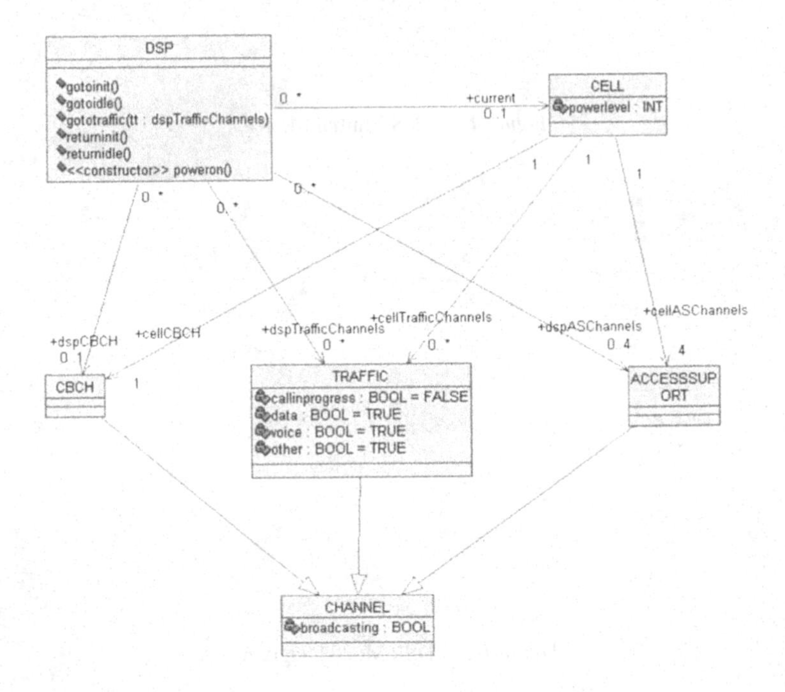

At this level of refinement an invariant was introduced to protect the DSP from being in certain modes without a correct set of channel configurations:

```
1| (
2|     (dsp_state=idle or dsp_state=traffic)=>
3|          dspCBCH=current.cellCBCH
4| ) &
5| (
6|     (dsp_state=idle or dsp_state=traffic) =>
7|          dspTrafficChannels<:current.cellTrafficChannels
8| ) &
9| (
10|    (dsp_state=idle or dsp_state=traffic)=>
11|         dspASChannels<:current.cellASChannels
12| ) &
13|   dspChannels=(dspCBCH  dspTrafficChannels  dspASChannels)
```

The most interesting feature explored here was that of how the inherited classes are constructed by utilising the - in this case - Channel class as the most abstract representation. This can be seen as being one instance of the notion of **decomposition** is the modelling methodology. More work is required to formalise this concept in this domain though.

Two points were investigated here, that of the disjunction of inherited classes and the replication of relationships. The first point is easily realised through the use of an invariant denoting that the three sets in this case are in fact disjoint. This invariant is shown below as a property of the generated B machine:

```
1| PROPERTIES
2|    CBCH / TRAFFIC={} &
3|    CBCH / ACCESSSUPORT={} &
4|    ACCESSSUPORT / TRAFFIC={}
```

Secondly we have the situation where we must (at present) manually decompose the relationship between Cell an Channel into three relationships between Cell and the three subclasses of Channel. This also then requires an invariant on the class Cell that glues the relationship cellChannels to that of the three new relationships for refinement purposes. This invariant is realised (on Cell) thus:

```
1| INVARIANT
2|    cellChannels = {cellCBCH}  cellTrafficChannels  cellASChannels
```

Note the singleton set for cellCBCH - this is part of the specification in this case as there is only a single cell CBCH instance to be considered here.

The second refinement as depicted in figure 1.4 in a continuation of that shown in 1.4 with further decomposition of the channel class structure through iheritance mechansisms.

Due to the nature of the development of the UML to B translations no further refinements were made with work concentrating mainly on the techniques for translating and developing the models.

One particular result warrants mentioning here and that is under some circumstances, especially those related to inheritance, the theorem prover as current implemented does fail with proof obligations regarding typing information.

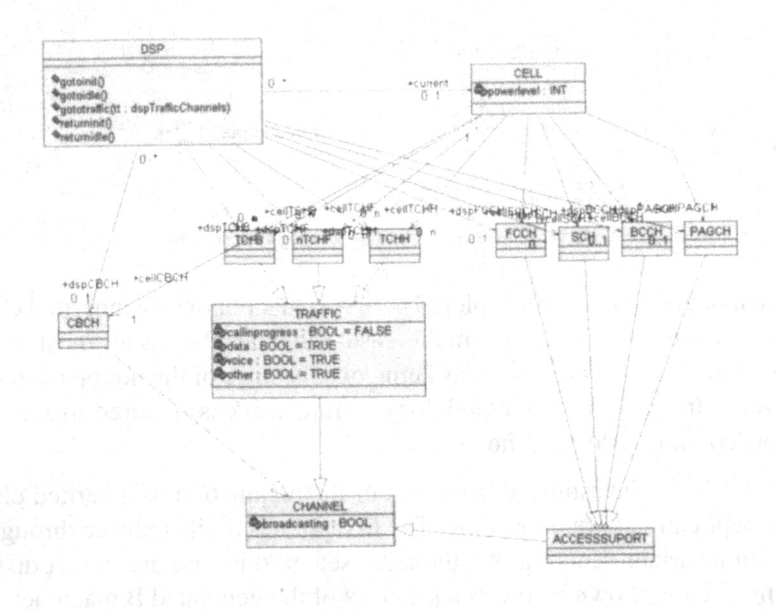

Figure 1.4. DSP Refined Model (2)

While these proof obligations are relatively trivial to solve more work is required on identifying and discharging those obligations.

1.3 Echo Cancellation Subsystem and Model Integration

Echo cancellation is a critical feature of systems where audio feedback is a possibility. The problem tackled in the design here is that of the audio crosstalk between what is broadcast on the speaker of a mobile device and what is picked up by the microphone. One wished to avoid the situation where one speaker hears what he/she said a few moments earlier being rebroadcast across the network.

The specification, design and implementation for this module is described in other PUSSEE work packages and also in the paper Krupp et al., 2004.

The architecture and the echo cancellation subsystem exist at two different levels of abstraction. It is a common situation where a number of subsystems, the architecture etc are being worked on concurrently and that integration of these parts proceeds as smoothly as possible with respect to their internal consistencies, the consistency of the system as a whole and certain global functional and non-functional properties.

Unfortunately within the PUSSEE project it has not been possible to proceed with a detailed analysis of how two subsystems are composed at this time.

2. THE PUSSEE METHOD (NOKIA)

For any techniques to work it is important to ensure that they can be integrated in with existing and know development strategies.

In Nokia, certainly for software development the Octopus Method is widely used despite it being considered old. However the concepts and methods discussed in Octopus are directly relevant to embedded systems and indeed this is the context in which it was developed.

As a contrast processes such as Ration Unified Process (RUP) are much more generic in nature and aimed at large scale, enterprise systems.

The PUSSEE Method from Nokia's perspective can be placed in the MDA framework such that we can equate the kinds of development approach favoured by OO principles and the B-method with development mappings and the UML to B and RIL translations through transformation mappings.

An overview of the method can be seen in figure 1.5 which shows how the modelling language and contents of those models changes over time and the relationship between those *working* models and the requirements that drive them and the models constructed for test purposes. We describe this method in more detail below.

The process overall is highly iterative and the above diagram should not be read as a process which only generates three models. In reality only one or two models made be produced in the core UML language, while the bulk of the remaining models will be made in the UML-B and the UML-B+RT+HW languages. Each model produced must be a refinement of the previous iterations that led to that model.

One of the sources of problems in the Nokia case studies has been the definition of the PUSSEE Method and in particular the notion of refinement and software development in the B-Method. However understanding the strengths and the weaknesses of the strict B-Method style versus Octopus/RUP/XP and how the methods can be integrated is of critical importance.

In this section we discuss the ways in which the PUSSEE process has been implemented, the problems encountered and the adopted solutions. This section deals with primarily the issues below a discussion of the tactics, such as refinement, used to develop the models through the development life cycle.

The PUSSEE Method is still very much in a developmental stage - this is unavoidable in such a short project where a number of techniques such as UML

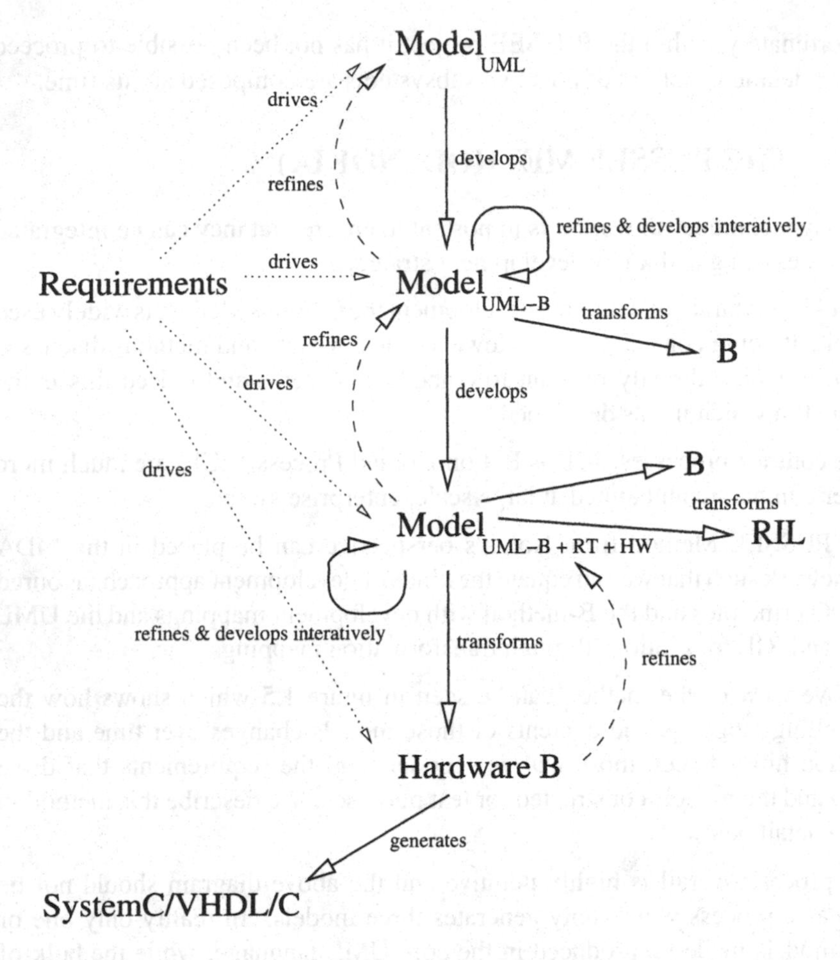

Figure 1.5. Method Overview

to B mappings have to be developed. However we feel that we have gained enough information for a strong basis in any continuation of the development of this particular method and the applications of the techniques here to other methods.

2.1 Overview of the MDA

The core ideas behind the MDA, that of all development is made through models and models are mapped to models and that everything is a model is not necessarily new. The 1970's and 80's ideas of CASE tool based developed embody these ideas but it is only now that we have the requisite technologies and experiences in place to make these concepts work in practise. A number of major problems exist in systems design, particularly we pick out the use of the UML and how the UML is used as being of primary concern here.

The UML is a (primarily) graphical specification notation with a weak semantics based upon object oriented concept. The problem here is that it is often not clear to the developer that he/she must tailor the UML and its semantics to the system being developed - the choice of semantics and notations here can be quite overwhelming.

From a methodological perspective it is often the case that modelling proceeds without respect to how models are developed but only to the process ("when"). Worse still the models developed are in one of two states, either a pictures of class diagrams without any real semantics and value, or containing implementation code in which case the semantics of the models is firmly rooted in that of the implementation language - normally C++ or Java. In both cases we have the problem that either the models lack value or that the models are so firmly bound to the implementation they only act as graphical versions of C++ or Java. Here we present the concepts behind the MDA and discuss the issues that the MDA raises with regards to the methodologies required to take advantage of this method of development.

The parimary concept in the MDA is that models are related by mappings. Current versions of the model describing this structure are exceeding (and probably with good reason) complex. Here we present a much simplified version of this structure which concentrates on the relationship between the concept of a model, the mappings and the language in which the model is written in.

The MDA is a complex structure which takes into consideration many aspects of modelling such as the language, semantics and model management. In figure 1.6 we show a simplified representation of the MDA meta-model written using UML.

From this model we can clearly see the separation of concerns provided by the MDA and embodied in the technologies on which it is based. For example the UML Uml, 2002 makes the distinction between notation and semantics. The UML is supplied with a weak semantics enough to suit nearly all development tasks and extensible enough for it to be customised to most domains.

The meta-model shown here we have reconstructed from our experiences in using the MDA and MDA-like approaches Oliver, 2002b. The primary issues are the creation of a structure of mappings and the explicit representation of the structure of a language. In the rest of this section we concentrate on the meaning of the two types of mapping presented here and their relationships via the mode to the underlying language. The final section of the paper discusses briefly the relationship and proposed ideas regarding methodologies and their semantics as expressed through the MDA.

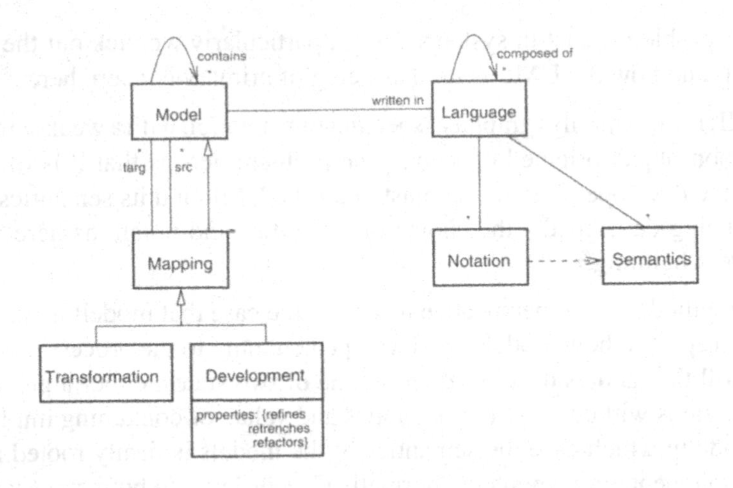

Figure 1.6. MDA Meta-Model

2.2 Mappings

The mapping is the fundamental construct of MDA. The key point about the mappings in the MDA is that they are of semantic nature and not syntactic nature - that is they map the concepts in one language to the concepts in another preserving the meaning. This in unlike the traditional syntactic mappings found in many tools, for example, those that map UML classes to C++ or Java classes - this is of course fine *if* the semantic gap between the diagram and the code is almost non-existent.

The MDA as it stands does not define any taxonomy of mappings, this we feel leads to some confusions about what a mapping is and what can be performed by a mapping.

In Oliver, 2002a is a description of the space in which a model exists known as the model matrix shown in figure 1.7. Here we can clearly see that a model exists in a many dimensional space corresponding to various aspects of the state and meaning of that model at any particular time. We concentrate here on just the 'vertical' and 'horizontal' axes to which we give the names

- Development or Vertical Mappings (shown in the model axis)
- Transformation or Horizontal Mappings (shown on the support axis)

We consider vertical mappings those that are conventionally in MDA parlance thought of as platform independent models (PIM) to platform specific models

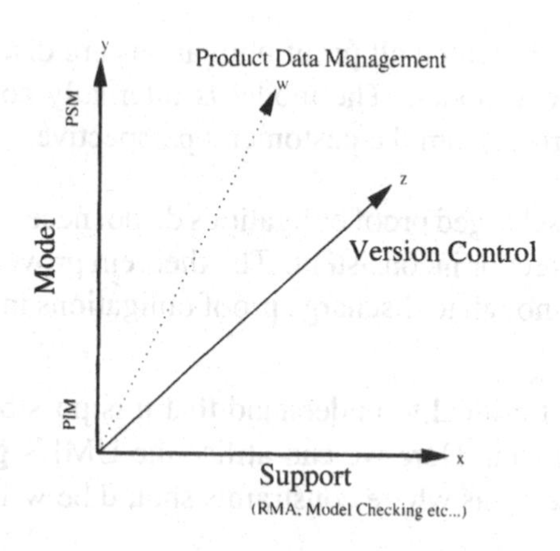

Figure 1.7. Model Matrix

(PSM) mappings, that is, those that change the abstraction level. Horizontal mappings we consider not to change the abstraction level.

3. FORMAL MODEL DEVELOPMENT

Model development in a formal manner within PUSSEE has a number of issues on how models are developed and in particular certain aspects of those models need to be investigated and modelled with a much greater degree of precision than would normally be necessary.

In reality these issues should be modelled but without any formal backup they are little more than just comments in the model that can not be checked in any way. The utilisation of B means that we have a means by which these issues can be checked. In addition if one does not deal with these issues then it is perfectly possible that a model and its generated code may be fully proved but admit situations that are illegal with respect to the intended properties.

In this section we discuss a number of ideas that have been proposed during this project. They are included here for completeness and for documentation. The ideas have not necessarily been widely disseminated nor implemented but represent possible future directions related to this work.

3.1 Proof Obligations and UML

The proof obligations generated from the B code give us information about the current integrity of the model. There are a couple of points which should be bared in mind here.

- Firstly a model in which all proof obligations are discharged is not necessarily a correct model. The model *is* internally consistent but is not necessarily correct from the customer's perspective.

- Secondly undischarged proof obligations do not necessarily mean that the model is incorrect or inconsistent. The theorem prover is not necessarily sophisticated enough to discharge proof obligations in all circumstances.

From the first point it critical to understand that it is possible that not enough constraints on the model. Here we can utilise the UML's graphical nature to assist in finding these areas where constraints should be written.

3.2 Identifying Constraints from UML

When writing class diagrams and in particular the relationships between classes and their attributes it is very easy to miss certain situations where constraints should be written but aren't. Here we describe some useful heuristics that can be applied to indentify such contraints

3.2.1 Attribute Pairs.

Any pair of attributes implies a constraint.

Every class contains a number of attributes (usually more than zero). It is the case that between every combination of attributes there exists a constraint - or in a simpler form, between any pair. The nature of the constraint however may be very weak such that the attributes are independent of each other no specific constraint need be written.

In figure 1.8 we can see a class with two attributes. From this we may supply an invariant linking these two attributes. An example invariant is shown below:

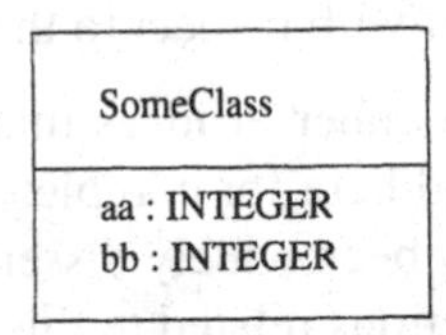

Figure 1.8. Example Attribute Pairs

```
1 INVARIANT
2     ...
3     aa > 1 => bb < 3
4     ...
```

3.2.2 Multiple Associations.

Any pair of classes that have two or more associations implies a constraint.

In figure 1.9 we can see a pair of classes with two relationships ss and rr between them. From this we may supply an invariant constraining these relationships such as to prevent sharing of objects. An example invariant is shown below:

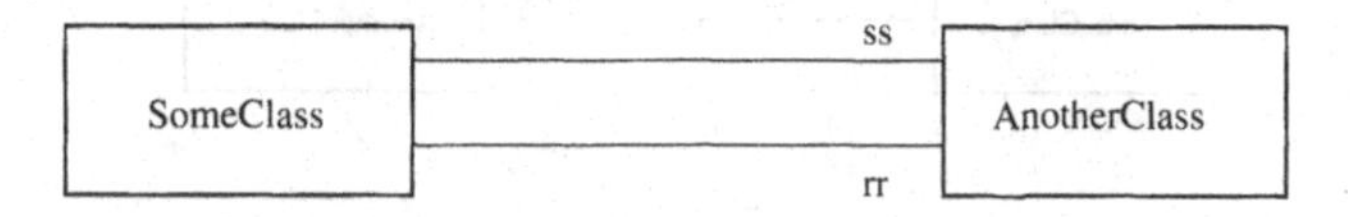

Figure 1.9. Example Multiple Associations

```
1 INVARIANT
2     ...
3     aa <: bb
4     ...
```

3.2.3 Shared Classes.

Any class that is shared implies a constraint

If any two classes share a third class through some kind of relationship then this implies some kind of constraint on the sharing of objects of that third class.

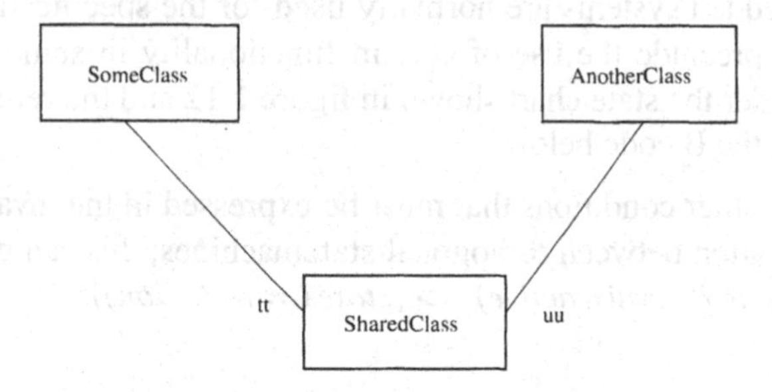

Figure 1.10. Example Shared Class

In figure 1.9 we can see a pair of classes with two relationships ss and rr between them. From this we may supply an invariant constraining these relationships such as to prevent sharing of objects. An example invariant is shown below:

```
1 INVARIANT
2     ...
3     not(ss=) => tt= &
4     not(tt=) => ss= &
5     ...
```

Note that this is a similar case to that described in §1.9.

3.2.4 Circularities.

> Any loop between three or more classes implies a constraint

This is the general case of the shared object scenario above and the invariant follows is a similar way.

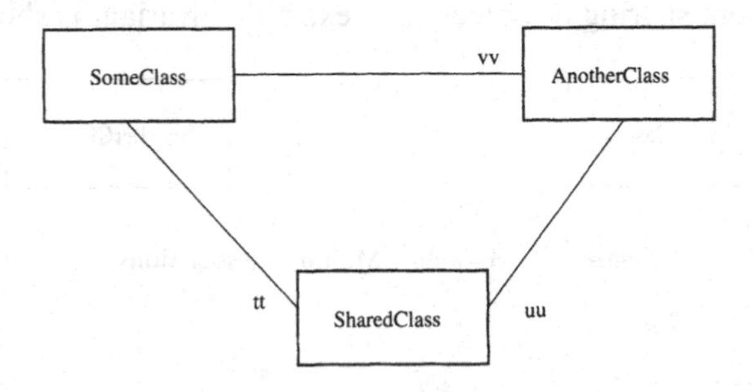

Figure 1.11. Example Class Circularity

3.2.5 States.

> A state implies an addition to the invariant and possibly also to the pre/post conditions of some operations.

States in embedded systems are normally used for the specification of modes and thus may preclude the use of certain functionality in some modes. For example consider the state chart shown in figure 1.12 and the references to the states given in the B code below.

Note there are other conditions that must be expressed in the invariant such as the synchronisation between orthogonal statemachines, this can be seen in the statement: *((state /= cruiseactive) => (state1 = noAction))*.

```
1  VARIABLES
2   state,
3   state1,
4   ...
5
6  SETS
7   STATE =  off, initialising, cruiseactive, cruisesuspended, shutdown  ;
8   STATE1 =  accelerating, braking, noAction
9
10 INVARIANT
11  ...
12  ((state /= cruiseactive) => (state1 = noAction)) &
13  ...
14
15
16 OPERATIONS
17
18 ...
19  periodicCheckSystem(currentSpeed) =
20    PRE
21      state = cruiseactive &
22      currentSpeed : NAT &   currentSpeed >= 0 &
23      currentSpeed <= 200
```

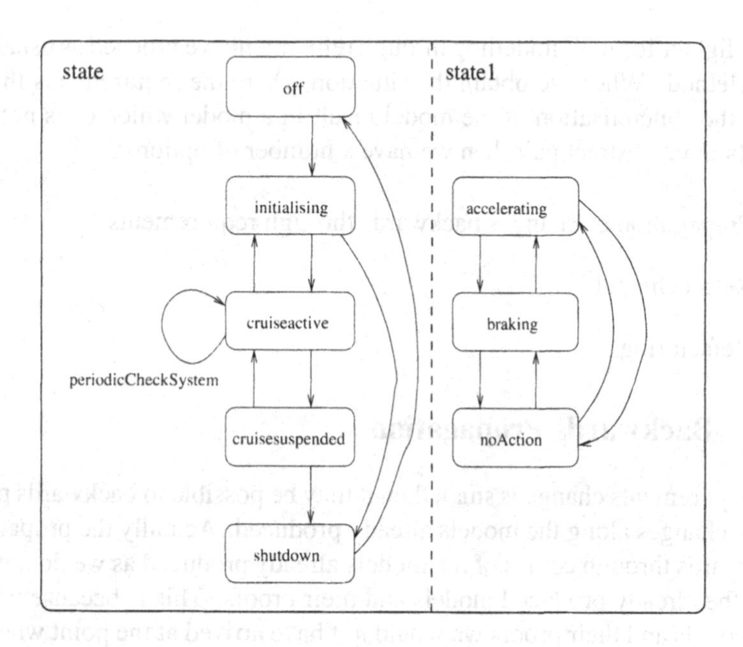

Figure 1.12. Example State Chart

```
24|   THEN
25|    SELECT
26|          (currentSpeed > requestedSpeed) THEN state1 := braking
27|     WHEN (currentSpeed = requestedSpeed) THEN state1 := noAction
28|     WHEN (currentSpeed < requestedSpeed) THEN state1 := accelerating
29|    END ||
30|     state := cruiseactive
31|   END;
32| ...
```

3.3 Refinement Difficulties

The notion of refinement in tightly coupled with that development process in the B-Method. Development of the software proceeds in a top-down fashion from abstract descriptions to more concrete descriptions and finally to a specification that can be implemented.

However in large scale software development and certainly in an industrial context requirements do change during the course of the software development life-cycle. It is also true to say that it is often the case that models and requirements develop in such a way that if it were not for the previous - now outdated requirements - then the current more accurate model and requirements would not exist. A very strong method such as B-Method does not allow this style of software development as embodied in most software development methodologies and processes today such as Octopus, RUP and Extreme Programming.

Using a lighter form of modelling through refinement, we proceed as usual with the B-Method. When we obtain the situation where the requirements that are driving the concretisation of the model result in a model which does not fully refine its more abstract pair then we have a number of options:

1 Propagation of changes backwards through requirements

2 Retrenchment

3 Refactoring

3.4 Backwards Propagation

If the requirements change is small then it may be possible to backwards propagate the changes along the models already produced. Actually the propagation is backwards through copies of the models already produced as we do not wish to lose the already produced models and their proofs. This is because without those models and their proofs we would not have arrived at the point where we would have found or needed the change in requirements, that is the previous models serve as historical reference.

Let us consider a system where we have proceeded through a number of refinement steps, eg: M0, M1 and M2. During the development of the next model, called M3, we find that M3 is not a refinement of M2 but generates a set of contradictions C. However the model M3 does indeed satisfy the current requirements and we can be confident that the model is correct because the proofs obtained from M0 through to M2 are correct and fully proven.

We now propagate the changes required back to a more abstract model M2' in such a way that M3 is a refinement of M2' and that M2' is not changed anymore than to satisfy the contradictions C. If one were to compare M2' and M2 one would see no more changes between these models than is necessary to support the M3-M2' refinement. In other words M2' is the model that we would have had if the requirements would have changed one step earlier. We believe that it is unlikely that one would want to propagate the changes back further, however we leave this possibility open.

The changes that would initiate or dictate using backwards propagation are relatively small but of the type where it is necessary to weaken the precondition of an operation and loosen the allowable values of some variable or type.

The formal support for this mechanism has been discussed but not fully developed. Investigation here has revealed retrenchment as a method that supports this issue but this has not been investigated in detail within the context of this project and is not supported by Atelier-B.

3.5 Retrenchment

Retrenchment Poppleton and Groves, 2003, Poppleton and Banach, 2002 is a generalisation of refinement in which the refinement step is supplemented by a set of concessions which weaken the refinement. These concessions give a formal specification of the requirements changes as well as providing a formal link to earlier, more abstract models.

Retrenchment must be used with case as it can be the case where the concessions can remove nearly or all formal links between the models. The concessions that are made upon a model form a lattice where the least concessions, ie: none, correspond to refinement.

M_1 retrenches M_0 with concessions C is written:

$$M_0 \lesssim_C M_1$$

Refinement can be expressed as follows:

$$M_0 \lesssim_\emptyset M_1$$

which is equivalent to writing $M_0 \sqsubseteq M_1$.
Finally the situation where we have

$$M_0 \lesssim_\perp M_1$$

is the situation where M_1 and M_0 have no conceptual link.
Currently within the formal methods community retrenchment is considered 'controversial' however it does provide at least a formal way of dealing with requirements change.
Within PUSSEE we have actively investigates retrenchment and its capabilities even though these are currently unsupported in B at the present time.

3.6 Refactoring

When the change in requirements is large it may be necessary to re-factor the model such that our model is now considered to be the top-level model or the most abstract model.

Similarly as before let us once again consider a system where we have proceeded through a number of refinement steps, eg: M0, M1 and M2. During the development of the next model, called M3, we find that M3 is not a refinement of M2 but generates a set of contradictions C. However the model M3 does indeed satisfy the current requirements and we can be confident that the model is correct because the proofs obtained from M0 through to M2 are correct and fully proven.

In this case the set of contradictions is too great to propagate backwards and it is necessary to 'forget' the previous models M0 through to M2 and their proofs and to consider M3 to be a new top-level model.

We do lose the previous proofs and the previous models no longer play an active role in the development of the system. Strict adherents to the B-Method may now even consider the re-factored model M3 to be too concrete a model. B-Method would insist however that at least one level of refinement more is made on the model before we can proceed to implementation and in this respect we feel that we are respecting the good features of the B-Method itself.

4. CONCLUSIONS

The problem with strict top-down development and strict refinement is that bottom up driven characteristics become apparent but can not be dealt with in a consistent manner. Here we have tried to take this into consideration.

While one may argue that we are deliberately breaking the strictness of B and thus losing some of the advantages it is also true that this strictness does impose some constraints that are unacceptable in *some kinds* of software development. In the authors' experiences it is true that requirements changes do happen and are inevitable in modern, large scale software development.

At the same time we do not wish to lose the advantages that we obtain from the B-Method or the B-language and in this respect we have tried to keep the best of both the formal and the more exploratory styles of programming without losing too much of the advantages and good practices of each.

At this point in time the suggested additions to B-method will probably not be implemented in any form other than comments.

Notes

1. Actually this was the idea here

References

(2002). *OMG Unified Modelling Language Specification)*. Object Management Group, version 1.5 edition. OMG Document Number ad/02-09-02.

Krupp, Alexander, Mueller, Wolfgang, and Oliver, Ian (2004). Formal refinement and model checking of an echo cancellation unit. In *Proceedings of Design, Automation and Test in Europe, DATE'04*. to be published.

Mouly, Michael and Pautet, Marie-Bernadette (1992). *The GSM System for Mobile Communicatons*. Cell and Sys. 9-782950-719003.

Oliver, Ian (2002a). Experiences of model driven architecture in real-time embedded systems. In *Proceedings of FDL02, Marseille, France, Sept 2002*.

Oliver, Ian (2002b). Model driven embedded systems. In Lilius, Johan, Balarin, Felice, and Machado, Ricardo J., editors, *Proceedings of Third International Conference on Application of Concurrency to System Design ACSD2003, Guimarães, Portugal*. IEEE Computer Society.

Poppleton, Michael and Banach, Richard (2002). Controlling control systems: an application of evolving retrenchment. volume 2272, pages 42–61. Springer-Verlag Lecture Notes in Computer Science.

Poppleton, Michael and Groves, Lindsay (2003). Software evolution with refinement and retrenchment. Department of Computer Science, Åbo Akademi University, Turku, Finland.

Chapter 16

UML-B SPECIFICATION AND HARDWARE IMPLEMENTATION OF A HAMMING CODER/DECODER

D. Cansell, S. Hallerstede, I. Oliver

Abstract: Formal refinement as offered by the B method has been shown to be applicable in practice and to scale up. However, it has been recognised that it is difficult communicate a formal B model with customers. Recently, the UML has been investigated as an interface to rigorous formal B models to facilitate this communication. The UML is generally accepted as being a good vehicle for communicating models of systems. Availability of this interface to the B method addresses a major problem faced by most formal methods: How to validate a formal model with a customer who is not formal methods expert?

In this chapter we present an approach to the development of a formally verified circuit implementing a Hamming encoder/decoder. The UML-B is used as a formal specification language and the B method is used to prove refinements until the implementation level at which we can translate into VHDL.
Key words: UML, EventB, error correction

1. INTRODUCTION

Most of the work that has been carried out in combining the UML and the B method has focused on the translation of UML models to B models. This has led to several good proposals on how to represent features of the UML in B, e.g. [19, 25]. By and large, we consider the problem of the translation as solved. We used the tool U2B [17] to translate UML-B models to B models and, more importantly, as interface to the B method. All specification was done in UML-B and refinements between UML-B models were proven in B. Our work has led to some new requirements on the

J. Mermet (ed.), UML-B Specification for Proven Embedded System Design, 261–277.
© 2004 *Kluwer Academic Publishers. Printed in the Netherlands.*

translation concerning easiness of proof. Until now, most translations have been made with exact representation of UML features in B in mind as measure of quality. We think that ease of proof is very important when the objective is to verify properties of a system formally. In B this verification is done by means of formal refinement. By using a translator from B to VHDL [18] from a B implementation of a circuit we have achieved a proved correct circuit by refinement in UML-B. This style of mapping between a UML profile and some other formalism fits very well with current trends towards the integration of 'model-based testing' techniques and Model Driven Architecture approaches [22, 21].

Our methodology is based on the event-based variant of the B method [7, 10, 11] which integrates the incremental development of formal models using a theorem prover to validate each step of development, called refinement. A model is simply defined as a reactive system with an invariant and safety properties that express requirements on the target System-on-Chip, together with hints on the architecture.

To demonstrate the approach we use a Hamming encoder/decoder. The system is specified and refined in EventB which is supported by U2B by attaching the « event system » stereotype to a model. The actual EventB specifications are not visible but represented by UML-B models. To perform refinement proofs an understanding of formal refinement remains essential, as well as, an understanding of the principles of EventB. We think, however, that the handling of actual specifications is easier in UML-B for people who are not experts in B. We have only anecdotal evidence of this though. In discussions about simple abstract specifications of a similar communication system we found that non-experts in formal methods could immediate join and help to improve the specification. In this context it was also found that animation and model-checking [20] were very useful in these interactive sessions. Proof was still carried out by experts though. Formal methods experts may value UML-B as a means to explain formal models to non-experts. The visual notation may serve as a common ground between experts in formal methods and experts in the application domains who have not been trained in formal methods. The general problem of the difficulties in creating formal models is pointed out in [27, 26].

We aim the production of a synchronous circuit design in VHDL. Implementations of circuit in B are expressed in a subset of the language, called BHDL, suitable for translation to VHDL. A similar approach is used for translations to programming languages like C. The subset of the B language used for software is called B0[13]. This is justified by the fact that plenty of synthesis software is commercially available for synchronous designs. By making use of synthesis software at the end of a B development one can say, a correct circuit has been derived (assuming that the synthesis software is correct and the model assumptions made in the B model are correct). The action system formalism [9] has been applied to the construction synchronous designs also [23]. Similarly to the B method, system development with action systems is based on refinement. In the action system approach synchronous designs are modeled using elements of an asynchronous model as primitives. We model synchronous circuits directly avoiding the extra complication in the case one is not interested in asynchronous properties.

2. EVENTB

2.1 Event-based modelling

Our event-driven approach [6, 2, 3] is based on the B notation [1]. It extends the methodological scope of basic concepts such as set-theoretical notations and generalised substitutions in order to take into account the idea of *formal models*. Roughly speaking, a formal model is characterised by a (finite) list x of *state variables* possibly modified by a (finite) list of *events*; an invariant $I(x)$ states some properties that must always be satisfied by the variables x and *maintained* by the activation of the events. Abstract models are close to guarded commands of Dijkstra [15], action systems of Back [8] and to UNITY programs [12]. In what follows, we briefly recall definitions and principles of formal models and explain how they can be managed by Atelier B [13].

2.2 Generalised substitutions

Generalised substitutions are borrowed from the B notation. They provide a way to express the transformations of the values of the state variables of a formal model. In its simple form, $x := E(x)$, a generalised substitution looks like an assignment statement. In this construct, x denotes a vector build on the set of state variables of the model, and $E(x)$ a vector of expressions of the same size as the vector x. The interpretation we shall give here to this statement is *not* however that of an assignment statement. We interpret it as a *logical simultaneous substitution* of each variable of the vector x by the corresponding expression of the vector $E(x)$.

2.3 Events and Before-After Predicates

An event is essentially made of two parts: a *guard*, which is a predicate built on the state variables, and an *action*, which is a generalised substitution. An event can take one of the forms shown in the table below. In these constructs, *evt* is an identifier: this is the event name. The first event is not guarded: it is thus always enabled. The guard of the other events, which states the necessary condition for these events to occur, is represented by $G(x)$ in the second case, and by $\exists t \cdot G(t, x)$ in the third one. The latter defines a non-deterministic event where t represents a vector of distinct local variables. The, so-called, before-after predicate $BA(x, x')$ associated with each event shape, describes the event as a logical predicate expressing the relationship linking the values of the state variables just before (x) and just after (x') the event "execution".

Event	BA-Predicate $BA(x, x_0)$
evt = begin $x : P(x_0, x)$ end	$P(x, x')$
evt = select $G(x)$ then $x : Q(x_0, x)$ end	$G(x) \wedge Q(x, x')$
evt = any t where $G(t, x)$ then $x : R(x_0, x, t)$ end	$\exists t \cdot (G(t, x) \wedge R(x, x', t))$

Proof obligations are produced from events in order to state that the invariant condition $I(x)$ is preserved. We next give the general rule to be proved. It follows immediately from the very definition of the before-after predicate, $BA(x, x')$ of each event: $I(x) \wedge BA(x, x') \Rightarrow I(x')$.

Notice that it follows from the two guarded forms of the events that this obligation is trivially discharged when the guard of the event is false. When it is the case, the event is said to be "disabled".

2.4 Model Refinement

The refinement of a formal model allows us to enrich a model in a *step by step* approach. Refinement provides a way to construct stronger invariants and also to add details in a model. It is also used to transform an abstract model in a more concrete version by modifying the state description. This is essentially done by extending the list of state variables (possibly suppressing some of them), by refining each abstract event into a corresponding concrete version, and by adding new events. The abstract state variables, x, and the concrete ones, y, are linked together by means of a, so-called, *gluing invariant* $J(x, y)$. A number of proof obligations ensure that (1) each abstract event is correctly refined by its corresponding concrete version, (2) each new event refines *skip*, (3) no new event take control for ever, and (4) relative deadlock-freeness is preserved.

2.5 Definition of Refinement

We suppose that an abstract model *AM* with variables x and invariant $I(x)$ is refined by a concrete model *CM* with variables y and gluing invariant $J(x, y)$. If $BAA(x, x')$ and $BAC(y, y')$ are respectively the abstract and concrete before-after predicates of the same event, we have to prove the following statement:

$$I(x) \wedge J(x, y) \wedge BAC(y, y') \Rightarrow \exists x' \cdot (BAA(x, x') \wedge J(x', y')).$$

This says that under the abstract invariant $I(x)$ and the concrete one $J(x, y)$, a concrete step $BAC(y, y')$ can be simulated ($\exists x'$) by an abstract one $BAA(x, x')$ in such a way that the gluing invariant $J(x', y')$ is preserved. A new event with before-after predicate $BA(y, y')$ must refine *skip* ($x' = x$). This leads to the following statement to prove:

$$I(x) \wedge J(x, y) \wedge BA(y, y') \Rightarrow J(x, y').$$

Moreover, we must prove that a variant $V(y)$ is decreased by each new event (this is to guarantee that an abstract step may occur). We have thus to prove the following for each new event with before-after predicate $BA(y, y')$:

$$I(x) \wedge J(x, y) \wedge BA(y, y') \Rightarrow V(y') < V(y).$$

Finally, we must prove that the concrete model does not introduce more deadlocks than the abstract one. This is formalised by means of the following proof obligation:

$$I(x) \wedge J(x, y) \wedge \text{grds}(AM) \Rightarrow \text{grds}(CM)$$

where grds(*AM*) stands for the disjunction of the guards of the events of the abstract model, and

grds(*CM*) stands for the disjunction of the guards of the events of the concrete one.

3. UML-B

The UML is a widespread modeling notation, popular among software and hardware engineers the like. The UML does not contain formal features that allows certain kinds of validation and verification of a model. The UML-B profile [28] has been developed to enable validation of models by animation, and verification though formal proof.

A subset of the UML consisting of packages, class diagrams, and state charts is combined with features of the EventB language to achieve graphical formal language. The translation from UML-B to EventB is automatic and can be done using the software tool U2B [17].

Certain stereotypes are used to provide a rich specification language and to constrain the syntax of UML-B models. We associate a package with the stereotype « event system » to indicate that the package represents an EventB model. Refinements are associated with the stereotype « refinement ». The UML is originally made for object-oriented modeling. We make little use of the object-orientation offered by the UML since EventB does not offer similar concepts. However, methods, attributes, and relations in UML-B are translated in an object-oriented way by associating them with objects of the classes they belong to. A class K is represented by a set K in B, and the objects are simply the elements k of K. Generally, all attributes $a : T$ of class K are translated to functions $a : K \rightarrow T$, and similarly for methods and relations. A stereotype « static » can be added to attributes, methods, and relations to change this behavior of the translator, so that $a : T$ is directly translated to $a : T$. Methods of UML-B correspond to events in EventB, and attributes and relations to variables. For more details on UML-B see [28].

To avoid overloading of UML-B diagrams, pieces of UML-B specifications can be written in a textual specification notation (similar to B) and attached to elements of the diagram. Some efforts have been made to give the B language embedded in UML-B an object-oriented flavor, mainly, by allowing a dot-notation for attribute-access customary in object-oriented languages. The high degree of formality of the B language remains though. Especially, good knowledge of first-order predicate logic and set theory is needed to use UML-B.

4. THE HAMMING SYSTEM IN UML-B

The abstract model *ABS* of the Hamming system simply copies a message from one buffer to another. There is no notion of coding and decoding. This is the most important property of the system. It should behave as if there were no errors in transmissions. If we implement this system by formal refinement we can be sure that an implementation we reach will effectively copy a message from a sender to a receiver. In between some computations may take place though. Transmission errors are introduced in a refinement of the abstract system, and we implement the Hamming code to compensate errors, finally leading to an implementation of a Hamming encoder and a Hamming decoder which are provably correct.

The refinement approach to system modeling offers the particular advantage of reducing the complexity of verification proofs. A system is initially specified by some important property. In the case of the Hamming system we say in the initial specification that there are no transmission errors. We then introduce transmission errors, and later we instantiate the coding and decoding of the Hamming codes, to prove that they eliminate transmission errors (if only one bit is corrupted). By proceeding in this stepwise manner, proof complexity is kept low in comparison to a single monolithic model of the implemented Hamming system. In this article we apply this method of system development to UML-B.

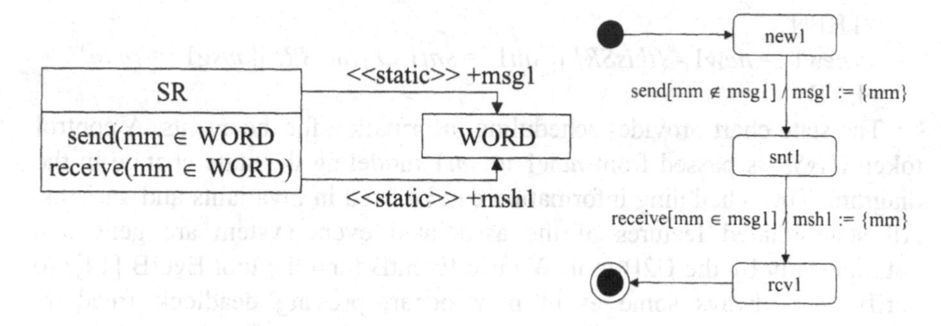

4.1 Abstract Model

Above we show the class diagram and state chart of the initial specification of the system in UML-B. A class *SR* models both sender and receiver of messages that are of type *WORD*. The multiplicity of *SR* has been set to 1 (this is not shown in the figure). It can be thought of as the process of sending and receiving. The message being sent is modeled by variable $msg1$, the message being received by variable $msh1$. The correctness property for reception is expressed by the invariant,

INVARIANT $msh1 \subseteq msg1$,

that expresses the fact that the message received must be equal to the message sent. We have also constrained the cardinality of $msg1$ in the invariant because we do so in later refinement steps in order to prevent overloading of diagrams. We feel that the use of UML is mostly to facilitate comprehension of models. Unwieldy diagrams would then appear to render the use of UML absurd. We use the following invariant to constrain $msg1$:

INVARIANT $\forall(m, n).(m \in msg1 \wedge n \in msg1 \Rightarrow m = n)$.

It says, that there is at most one element in $msg1$, in other words, if there are two elements we may pick from $msg1$ they must be identical. The system has two events *send* and *receive* that describe its possible evolutions. The parameterization of these events corresponds to the selection of corresponding variables in ANY-statements and is convenient for reference in state charts (see figure above). The behavior of the system is described by the state chart. Being in state $new1$ a message mm may be sent and copied into set $msg1$. On reception the messages the message is copied into set $msh1$. States are translated into disjoint sets that form a partition of SR. The condition $thisSR \in ss$ means that $thisSR$ is in state ss. Note, that in the Hamming system $thisSR \in ss$ is equivalent to $\{thisSR\} = SR$. For the state chart in the figure above we get:

INVARIANT $(new1 \subseteq SR \wedge ...) \wedge (new1 \cap snt1 = \varnothing \wedge ...)$.

This is the event corresponding to *send*:

$send =$ ANY $thisSR, mm$ WHERE
  $thisSR \in new1 \wedge mm\ 2\ WORD \wedge msg1 = \varnothing$
THEN
  $new1 := new1 - \{thisSR\} \mid\mid snt1 := snt1 \cup \{thisSR\} \mid\mid msg1 := \{mm\}$
END

The state chart provides scheduling information for the events. A control token *thisSR* is passed from $new1$ to $snt1$ modeling the state change in the diagram. The scheduling information can be used in invariants and variants. All state related features of the associated event system are generated automatically by the U2B tool. We use EventB (and the tool Evt2B [14]) to verify that always some event may occur, proving deadlock freedom. Deadlock-freedom is proved modulo $msh1 \neq \varnothing$, the termination condition of the system. It is necessary to prove deadlock-freedom to prevent the introduction of unwanted deadlocks. Otherwise, we would not be able to guarantee that the Hamming system implementation would actually compute anything at all (but just stop). After having proved the proof obligations generated by AtelierB [13, 5] we know also that the invariant holds, and thus the system is correct.

4.2 Refined Model

Starting from the initial abstract system in Figure 2 we traverse a sequence of refinements until we reach the implementation level where we can use a translator to generate VHDL describing the circuit that implements the Hamming encoder/decoder. Below we present the first refinement *REF* of abstract system *ABS*.

As we progress in the refinement process we add a lot of detail and UML diagrams tend to get overloaded, so that we prefer to write properties of the system in invariant clauses. As we progress refining we mostly represent data structures and state transitions by UML diagrams. For data manipulation and mathematical properties we use B to a larger degree. In

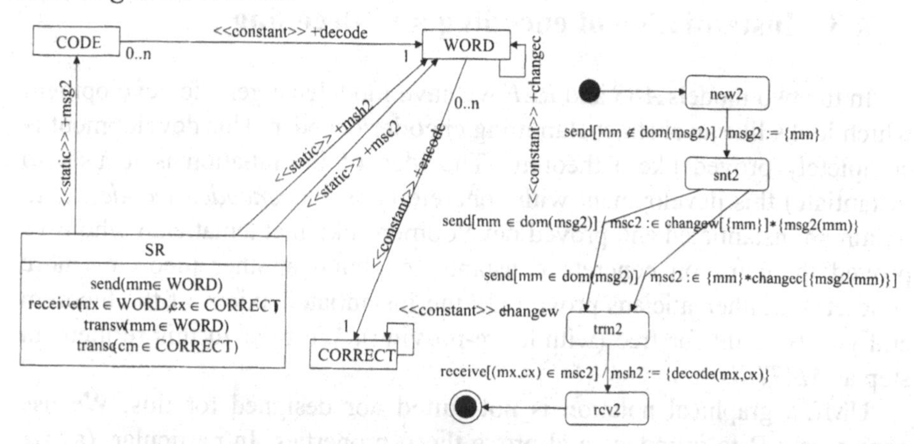

REF the sender sends a correction code with each transmitted word. Variable *msg2* is a subset of *CODE*. We have put a clause

INSTANCES *WORD * CORRECT*

into class *CODE* to express that an actual code sent consists of the original word plus a correction code.

This refinement step introduces possible transmission errors in the system. We express this by the new events *transc* and *transw*. Details about the transmissions errors are specified by properties in the B language, e.g.:

$$\forall(m,x).(m \in WORD \wedge x \in changec[\{encode(m)\}] \Rightarrow decode(m,x) = m) \wedge$$
$$\forall(m,n).(m \in WORD \wedge n \in changew[\{m\}] \Rightarrow decode(n,encode(m)) = m) \quad (1)$$

As expressed in the UML-B diagram, *encode* is a total function from *WORD* to *CORRECT* and *decode* is a total function from *CODE* to *WORD*. These algebraic properties of *encode* and *decode* are required for the refinement proof. They also state in simple mathematical terms the behaviour of the system consisting of an encoder and a decoder. A gluing invariant is used link the states of the two systems to prove the refinement relationship. One part of it is responsible for the states

INVARIANT $new2 \subseteq SR \wedge new1 = new2 \wedge rcv1 = trm2 \cup rcv2 \wedge \ldots$,

another one, that must be supplied by the user, for the data:

INVARIANT $msg1 = dom(msg2) \wedge decode[msc2] \subseteq msg1 \wedge \ldots$.

Because we always send the correct message as the first component of the code in *msg* we can recover the original message *msg*1 from the domain of *msg*2, that is $msg1 = dom(msg2)$. Variable *msc*2 contains the possibly corrupted message. In fact, it contains all possible corruptions and the uncorrupted message; but we do not whether the message that is being received is uncorrupted. The item $decode[msc2] \subseteq msg1$ of the invariant says that, if we decode any message of *msc*1 we always get the message that was originally sent.

Had we used identical names for states *new*1 and *new*2 (or *msh*1 and *msh*2) Atelier B would have taken care of the equality in the gluing invariant after renaming them appropriately. We have not used this feature of Atelier B for didactic reasons.

4.3 Instantiation of encoding and decoding

In the two models *ABS* and *REF* we have modeled a generic development which looks like an abstract Hamming encoder/decoder. This development is completely proved like a theorem. The idea of instantiation is to use (to instantiate) this development with concrete constants *encode*, *decode*, ... to obtain an instantiated and proved development like mathematician who uses proved theorems on concrete constants to obtain another theorem (more concrete). Mathematicians prove only the instantiated axiom of the theorem and get its result for free (without re-proving). We refer to this refinement step as *MATHS*.

UML's graphical notation is not suited nor designed for this. We use exclusively B to introduce and prove these properties. In particular, (a) we have to specify what kind of error may occur in a transmission, and (b) we must prove that the Hamming code is suitable to correct these errors. We deal with (1) by specifying details about the relations *changec* and *changew*, for instance, that there is at most one corrupted bit in a message:

$\forall\ (m,n).((m,n)\in changew \Rightarrow \forall\ (x,y).(m(x)\neq n(x) \wedge m(y)\neq n(y) \Rightarrow x=y))$.

Problem (b) is solved by specifying details about *encode* and *decode* and proving that property (1) holds. This property is now considered a theorem that we have to prove with respect to the instantiated constants *changew*, *changec*, *encode* and *decode*. We also have to precise the data-types *WORD* and *CORRECT* to be able to carry out proofs involving concrete messages of type *WORD*. We let $WORD = 1..4 \rightarrow 0..1$ and $CORRECT = 1..3 \rightarrow 0..1$.

From here we proceed as before by writing refinements in UML-B by introducing new states and events. This process is continued until a refinement is reached that is implementable. To arrive at the hardware controller for encoding and decoding we must recompose the corresponding events that have been introduced during the refinement process. Re-composition is discussed in a later section.

4.4 Evaluation

State charts are translated to sets that provide scheduling information that is useful for stating variants. The variant that is required to prove absence of live-lock in the first refinement is simply card(*snt2*). The U2B tool could generate some variants and remind the user to specify those, that cannot be easily generated, for instance, if there is a loop consisting only of new events. U2B does not generate this information at the moment. Besides, the states are a convenient representation for scheduling information that help to make the specification more comprehensible. Possible orders of events in event systems are often difficult recognise without them.

States of UML-B add sometimes redundancy to a specification. We could easily specify the initial system using variables *msg*1 and *msh*1 only. On the other hand the difficulty of the proofs does not increase. Only the number of proof obligations is higher. The table below shows the proof effort when using either EventB on its own, or in conjunction with UML-B.

	ABS		REF		MATHS	
	#PO	#IP	#PO	#IP	#PO	#IP
EventB	7	2	17	7	10	8
UML-B	20	4	40	14	10	8

The three columns in the table denote the abstract system *ABS*, its first refinement *REF*, and the refinement providing mathematical properties of Hamming codes *MATHS*, respectively. The number of generated proof obligations #PO and the number of proof obligations to be proved interactively #PI is displayed. Since we have not used UMLB to specify mathematical properties the proof effort for both cases is equal. For *ABS* and *REF* the number of proof obligations generated for the UML-B approach is considerable higher. However, many of them are discharged automatically by Atelier B. Many of those remaining to prove interactively can be discharged easily using a single interactive prover command. These proof obligations have the form:

$$new1 \cup snt1 \cup rcv1 = SR \Rightarrow (new1 - \{thisSR\}) \cup (snt1 \cup \{thisSR\}) \cup rcv1 = SR .$$

Corresponding proof scripts could be generated automatically, or Atelier B could be instructed to try the appropriate method on all proof obligations that were not discharged (as displayed in the table above). For *ABS* this would remove 2 proof obligations, and for *REF* it would remove 7. So the proof effort would be equal. The remaining proofs are of similar difficulty whether using EventB on its own, or using UML-B. To prove deadlock-freedom in the UML-B we needed to add clauses to the invariant linking states to guards of events. This has already been taken account of in the figures presented above. The clauses seem obvious but are not generated automatically by U2B:

$$(msg1 \neq \varnothing \Rightarrow snt1 \cup rcv1 = SR) \wedge (msh1 = \varnothing \Rightarrow rcv1 = \varnothing) .$$

The singleton set *SR* for modeling the system component seems redundant, however it is easy to understand, to explain, and does not add to proof complexity.

The method of constraining the states through the attributes is contrary or counter-intuitive to 'traditional' reactive system modeling techniques [24, 16]. This method has the advantage that we can prove live-ness properties of the system easily and simplifies the proof obligations generated.

4.5 Re-composition

Eventually we reach a refinement *IMP* at an abstraction level that we consider an implementation. Event system *IMP* consists of a collection of events that do not immediately correspond to entities that can be translated into software or hardware. We need to bring the system into a suitable form first. Fortunately, most of this can be carried out automatically. We note that in a event system we have different kinds of event, some belong to the controller that we intended to develop, and some to the environment in which it will run. The implementation contains a model of the acceptable behaviour of the environment, specifying transmission errors for which the controller works correctly. In system *IMP* we find three groups of events:

send	hamming encoder
error, correctY, correctN, receive	hamming decoder
transw, transc	environment

Event *send* contains everything we need to implement a hamming encoder. On the contrary, the decoder is split over four events *error*, *correctY*, *correctN*, *receive*. Event *error* computes the position of an error in a message, *correctY* corrects an error if there is one, and *correctN* deals with the case of an uncorrupted message.

We proceed in two steps. First the four events for the decoder are recomposed into a single event. Afterwards, we factor out the part of the events that constitutes the controllers for encoding and decoding messages. The first step is automatic (we have not implemented a re-composition tool yet though). The second can be performed by copying and pasting into a BHDL machine that is subsequently imported.

Re-composition happens by application of a number of re-composition rules. An events system is a collection of guarded events on which these rules applied. We only present the if-introduction rule[4]:

$$\frac{\text{SELECT } P \wedge Q \text{ THEN } S \text{ END [] SELECT } P \wedge \neg Q \text{ THEN } T \text{ END [] } U}{\text{SELECT } P \text{ THEN } \mathbf{IF}\ Q\ \mathbf{THEN}\ S\ \mathbf{ELSE}\ T\ \mathbf{END} \text{ END [] } U}$$

where [] denotes non-deterministic choice; P and Q denote predicates, and S, T, and U substitutions. In the context of UML-B, predicate P mainly handles the part of the guards dealing with states, e.g. $thisSR \in new1$ in event *send.*

In the next step we create a BHDL machine as shown below (2). These may be imported into an event system to prove them correct with respect to the system that we have specified. Next we present shows event *send* of the implementation of the Hamming code system after re-composition and factoring:

SYSTEM *MOD*
IMPORTS *HAMC, ...*

VARIABLES $msg4 \subseteq CODE \wedge ...$
EVENTS
$send$ = ANY $thisSR, mm$ WHERE
$mm \longleftarrow HAMC(cd);$
$msg4 := \{(mm,cd)\} \parallel ...$
END;
...
END

The notation $mm \longleftarrow HAMC(cd)$ denotes invocation of the operation of the BHDL machine *HAMC*. Semantically this is defined analogously to procedure calls in the B language.

4.6 Hardware Implementation

After the events that correspond to the encoder and the decoder have been recomposed respectively, a BHDL machine is factored out that contains the hardware description. In principle, this machine could be refined further using the B method. We consider it as an implementation here though. The BHDL implementation can be translated into VHDL by the BHDL translator. A BHDL machine may contain a subset of the B language suitable for translation to hardware. Similar sub-languages and corresponding translators exist for software, e.g. a translator to C.

This is the BHDL machine that calculates the correction code for a message of type *WORD*:

MACHINE *HAMC* (2)
SEES *TYPES*
INPUT *WRD*
OUTPUT *CCT*
INVARIANT
$WRD \in WORD \wedge CCT \in CORRECT$
OPERATION $CCT :=$
$\lambda kk.(kk = 1 | (WRD(2) + WRD(3) + WRD(4)) \bmod 2) \cup$
$\lambda kk.(kk = 2 | (WRD(1) + WRD(3) + WRD(4)) \bmod 2) \cup$
$\lambda kk.(kk = 3 | (WRD(1) + WRD(2) + WRD(4)) \bmod 2)$
END

The type declaration have been put into a machine called *TYPES*. The translator requires this in order to produce proper VHDL. It produces a VHDL package containing the type declarations. Machine *HAMC* is fairly simple and corresponds directly to the combinatorial VHDL design:

```
entity HAMC is port (
        WRD : in WORD;
        CCT : out CORRECT);
end;
architecture BEH of HAMC is begin
COMB : block begin
        CCT(1) ⇐ (((WRD(2) + WRD(3)) + WRD(4)) mod 2);
        CCT(2) ⇐ (((WRD(1) + WRD(3)) + WRD(4)) mod 2);
        CCT(3) ⇐ (((WRD(1) + WRD(2)) + WRD(4)) mod 2);
end block;
end;
```

For even only slightly larger BHDL machines the corresponding VHDL design is considerably more complicated and larger. The produced VHDL is mostly synthesisable. Some non-synthesisable features are present to allow for efficient simulation. Refinement can be used to remove these so as to arrive at an entirely synthesisable model [29]. When a synthesisable design is produced we have a achieved a proven correct hardware implementation. See the remarks in the next paragraph on the correctness of the generated VHDL.

The translator is proven correct with respect to an abstract representation of a set of hardware description languages. The abstract representation is mapped to the different target languages, like VHDL or SystemC. Since neither VHDL nor SystemC have a formal semantics the correctness proof for the translation is not possible directly. To avoid ambiguities we have restricted the constructs that can be used in the target languages.

5. CONCLUSION

The UML-B profile makes it possible to work with an integrated specification/modeling style of visual elements and formal notation. We consider the B model that results from translation to be semantically isomorphic to the UML model. As a consequence we can use formal techniques to prove properties about the system being modeled. In particular B-refinement can be used to prove properties of more complex systems.

We have presented how UML-B can be used with B refinement. Having a visual interface to

the B-method is appealing to people without special knowledge of formal methods and assists in the system size scalability problem encountered with some formal specifications. To them the more familiar notation of UML open the possibility to write formal specifications. To expert in formal methods the visual notation of the UML is convenient to explain formal models to non-experts. Much of the B formalism remains visible however. This is the case when specifying properties of data elements of a system, but especially when dealing with parts of a specification that are inherently of mathematical nature. This goes to a point when at particular steps we work exclusively in the B notation. This does not really appear surprising, because at some point one will always have to develop a theory underlying the modeling domain.

We found that although the amount of proofs increases by using UML-B, this does not cause an increase in the real effort that has to be put into proof. The additional proof obligations created can be trivially discharged.

Finally, we have presented a complete development that begins with a visual specification in UMLB and ends with a hardware design in VHDL. We have refined this initial model in UML-B using the B-method, proving that each model in the chain has all the properties of its successors. Thus we have created a provably correct circuit of a UML-B model.

6. REFERENCES

[1] J.-R. Abrial. *The B-Book - Assigning Programs to Meanings*. 1996.

[2] J.-R. Abrial. B#: Toward a synthesis between z and b. In D. Bert and M. Walden, editors, *3nd International Conference of B and Z Users - ZB 2003, Turku, Finland*, Lectures Notes in Computer Science, June 2003.

[3] J.-R. Abrial and L. Mussat. Introducing Dynamic Constraints in B. In D. Bert, editor, *B'98: Recent Advances in the Development and Use of the B-Method*, volume 1393 of *LNCS*, pages 83–128, 1998.

[4] Jean-Raymond Abrial. Event Driven Sequential Program Construction, 2000. http://www.matisse.qinetiq.com/links.htm.

[5] Jean-Raymond Abrial and D. Cansell. Click'n'prove: Interactive proofs within set theory. In David Basin and Burkhart Wolff, editors, *16th Intl. Conf. Theorem Proving in Higher Order Logics (TPHOLs'2003)*, volume 2758 of *Lecture Notes in Computer Science*, pages 1–24. Springer Verlag, September 2003.

[6] J.R. Abrial. *Extending B without changing it (for developing distributed systems)*. In H. Habrias, editor, *1st Conference on the B method*, 11 1996.

[7] J.R. Abrial. Formal construction of proved circuits. Technical report, 08 2001. Internal Note.

[8] R. J. R. Back. On correct refinement of programs. *Journal of Computer and System Sciences*, 23(1):49–68, 1979.

[9] R. J. R. Back and J. von Wright. Trace refinement of action systems.

In Bengt Jonsson and Joachim Parrow, editors, *CONCUR '94: Concurrency Theory, 5th International Conference*, volume 836 of *Lecture Notes in Computer Science*, pages 367–384, Uppsala, Sweden, 1994. Springer-Verlag.

[10] D. Cansell and D. M´ery. Integration of the proof process in the system development through refinement steps. In Eugenio Villar, editor, *5th Forum on Specification & Design Language - Workshop SFP in FDL'02 , Marseille, France*, Sep 2002.

[11] D. Cansell, Camel Tanougast, Yves Berviller, D. M´ery, Cyril Proch, Hassan Rabah, and Serge

Weber. Proof-based design of a microelectronic architecture for mpeg-2 bit-rate measurement. In

Forum on specification and Design Languages - FDL'03, Frankfurt, Germany, Sep 2003.

[12] K. M. Chandy and J. Misra. *Parallel Program Design A Foundation*. Addison-Wesley Publishing

Company, 1988. ISBN 0-201-05866-9.

[13] ClearSy – Systems Engineering, Aix-en-Provence, France. *Atelier B*. Software.

[14] ClearSy – Systems Engineering, http://www.atelierb.societe.com/documents.htm. *evt2B*. Software.

[15] E. W. Dijkstra. *A Discipline of Programming*. Prenctice-Hall, 1976.

[16] Bruce Powel Douglas. *Doing Hard Time. Developing Real-Time Systems with UML, Objects, Frameworks and Patterns*. Addison-Wesley, 1999. 0-201-49837-5.

[17] ECS, University of Southampton, http://www.ecs.soton.ac.uk/cfs/U2Bdownloads/U2Bdownloads.htm. *U2B Tool*. Software.

[18] KeesDA S.A., Grenoble, France. *BHDL User Guide*.

[19] Hung Ledang and Jeanine Souqui`eres. Contributions for modelling UML state-charts in B. *Lecture Notes in Computer Science*, 2335:109–??, 2002.

[20] Michael Leuschel and Michael Butler. ProB: A model checker for B. In Keijiro Araki, Stefania Gnesi, and Dino Mandrioli, editors, *FME 2003: Formal Methods*, LNCS 2805, pages 855–874. Springer-Verlag, 2003.

[21] Ian Oliver. Experiences of model driven architecture in real-time embedded systems. In *Proceedings of FDL02, Marseille, France, Sept 2002.*, 2002.

[22] Ian Oliver. Model driven embedded systems. In Johan Lilius, Felice Balarin, and Ricardo J. Machado, editors, *Proceedings of Third International Conference on Application of Concurrency to System Design ACSD2003, Guimar˜aes, Portugal*. IEEE Computer Society, June 2002.

[23] Juha Plosila and Tiberiu Seceleanu. Synchronous action systems. Technical Report TUCS TR-192, Turku Centre for Computer Science, Finland, 1998.

[24] Bran Selic, Garth Gullekson, and Paul T. Ward. *Real-Time Object-Oriented Modelling*. Wiley, 1994.

[25] Colin Snook. Combining UML and B. In *Proceedings of FDL'02*, 2002.

[26] Colin Snook and Michael Butler. Using a Graphical Design Tool for Formal Specification. In Dumke and Abran, editors, *Proceedings 13th Annual Workshop of the Psychology of Programming Interest Group*, 2001.

[27] Colin Snook and Rachel Harrison. Practitioners' Views on the Use of Formal Methods: An Industrial Survey by Structured Interview. *Information and Software Technology*, 43(4):275–283, 2001.

[28] Colin Snook, Ian Oliver, and Michael Butler. Towards a UML Profile for UML-B. Technical report, University of Southampton, 2003.

[29] VSI Working Group. IEEE P1076.6/D2.01 – Draft Standard For VHDL Register Transfer Level Synthesis. Unapproved Draft, IEEE, 2001.

Chapter 17

THE PUSSEE METHOD IN PRACTICE

Specification and design of an embedded telecom system based on HIPERLAN/2 protocol

Nikolaos S. Voros
INTRACOM S.A., Patra, Greece

Abstract: The main goal of this chapter is to demonstrate the feasibility of PUSSEE development framework. For that purpose, a case study borrowed from the telecommunication domain is used in order to exhibit the applicability the method and associated tools for the design of complex systems. The application described is part of an embedded system based on HIPERLAN/2 protocol.

Key words: UML-B profile, embedded telecom systems, formal refinement, system decomposition.

1. AN OVERVIEW OF HIPERLAN/2 PROTOCOL

HIPERLAN/2 protocol provides data rates up to 54 Mbits/sec for short range (up to 150 m) communications in indoor and outdoor environments. Typical application environments are offices, homes, exhibition halls, airports, train stations and so on.

In order to specify a radio access network that can be used with a variety of core networks, the HIPERLAN/2 standard [1] provides a flexible architecture that defines core independent physical (PHY) and Data Link Control (DLC) layers and a set of convergence layers that facilitate access to various core networks including Ethernet, ATM and IEEE 1394 (Firewire).

The air interface is based on time division duplex (TDD) and dynamic time division multiple access (TDMA). It relies on cellular networking topology combined with ad-hoc networking capability, and supports two basic modes of operation: centralized mode (CM) and direct mode (DM). In

J. Mermet (ed.), UML-B Specification for Proven Embedded System Design, 279–292.
© 2004 *Kluwer Academic Publishers. Printed in the Netherlands.*

the CM operation every radio cell is controlled by an access point covering a certain geographical area, and mobile terminals communicate with one another or with the core network through the access point. In the DM operation, mobile terminals in a single cell network can exchange data directly with one another. The access point controls the assignment of radio resources to the mobile terminals. Figure 17-1 outlines the protocol architecture, while Figure 17-2 delineates the scope of HIPERLAN/2 standards.

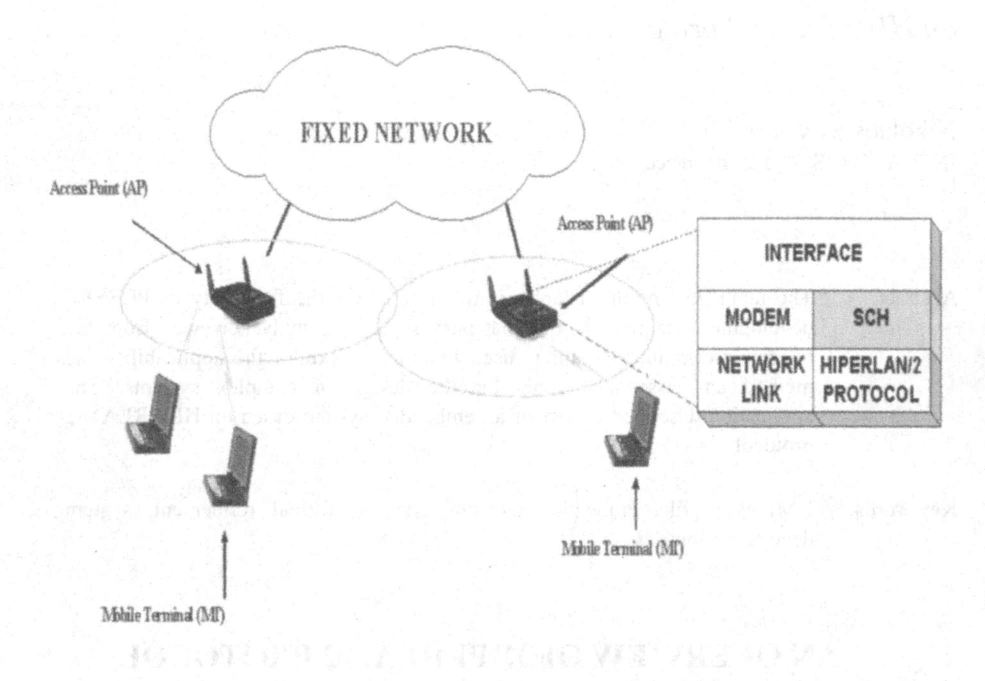

Figure 17-1. An overview of HIPERLAN/2 architecture

The system under design is part of the access point system and consists of the AP scheduler and the modem. The next paragraphs describe the design of the specific case study using the design steps supported by PUSSEE method.

In Figure 17-3 the final architecture of a prototype for the HIPERLAN/2 based system is described. The final system implementation employs both hardware (e.g. the HIPERLAN/2 modem), and software (e.g. DLC layer) components. Regarding the software part of the system, the design of the Frame Scheduler for the Access Point (AP) is presented. The latter, lies in the MAC sub layer of the DLC layer and is responsible for the design of MAC frames.

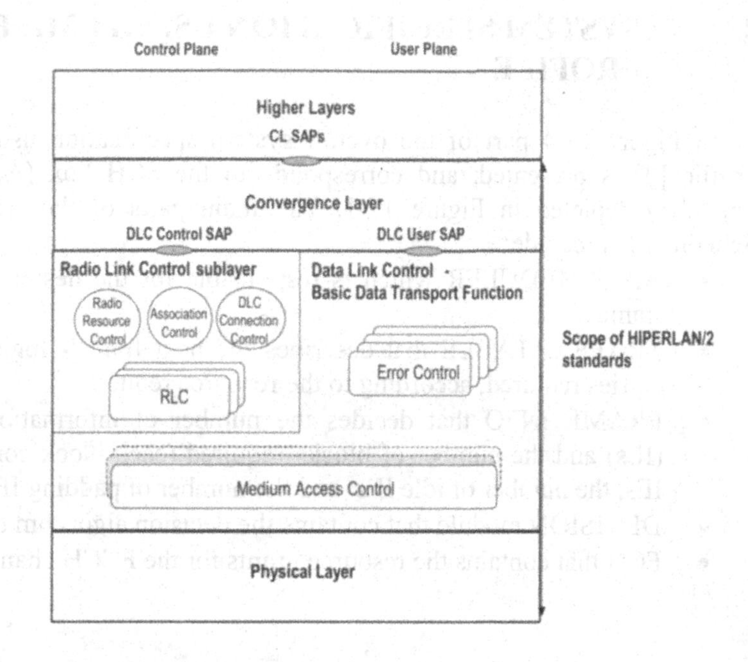

Figure 17-2. The scope of HIPERLAN/2 standards

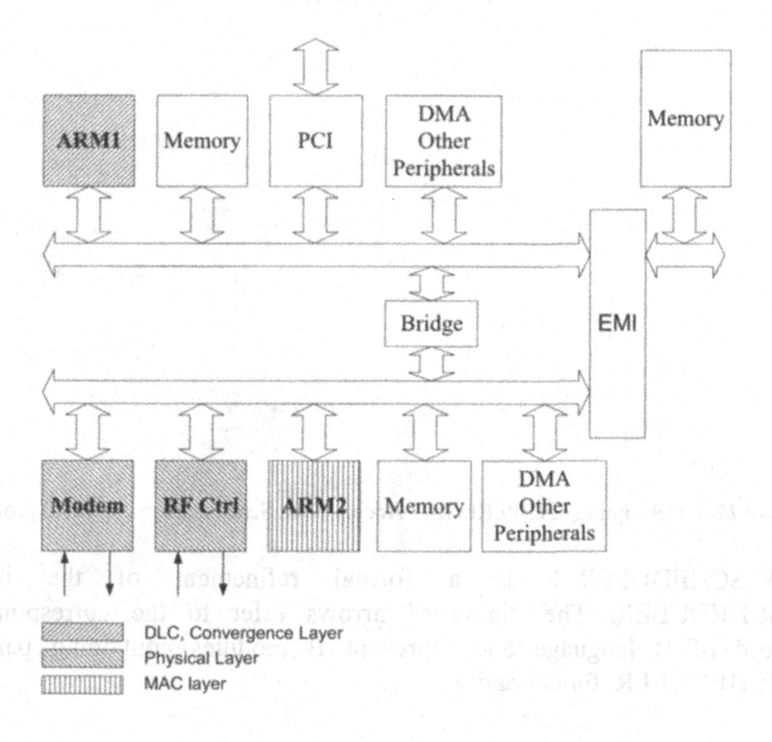

Figure 17-3. The HIPERLAN/2 prototype

2. SYSTEM SPECIFICATION USING UML-B PROFILE

In Figure 17-4 part of the overall system specification using UML-B profile [2] is presented, and corresponds to the SCH box (Access Point Sceduler) depicted in Figure 17-1. The main parts of the Access Point Scheduler [1] include:

- AP_SCHEDULER which is responsible for the design of a MAC frame.
- TRAFFIC_TABLE that describes the next frame's logical channel entries required, according to the resource requests.
- FRAME_INFO that decides the number of information elements (IEs) and the number of blocks required (each block contains three IEs, the number of idle IEs and the number of padding IEs).
- DECISION module that contains the decision algorithm used.
- FCH that contains the resource grants for the FCCH channel.

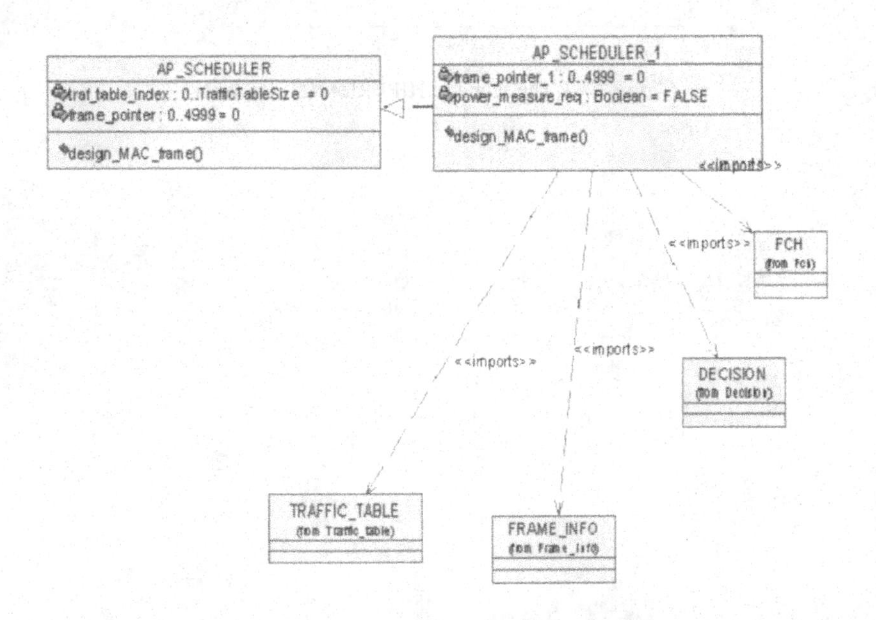

Figure 17-4. Description of HIPERLAN/2 Access Point Scheduler using UML-B profile

AP_SCHEDULER_1 is a formal refinement of the initial AP_SCHEDULER. The "imports" arrows refer to the corresponding keyword of B language and represent B modules containing part of AP_SCHEDULER functionality.

3. FORMALLY PROVEN MODEL REFINEMENT

For the design of the AP scheduler subsystem, the UML models of Figure 17-4 were translated to B machines using U2B translator [3]. The functionality of each class (including its attributes and operations) is thoroughly described in [4].

Figure 17-5 presents the B code produced by U2B translator for the AP_SCHEDULER class. AP_SCHEDULER_1 class was also translated to B using U2B, and the required proof obligations for the specific model refinement were generated and proven using Atelier B [5], as depicted in Figure 17-6.

```
MACHINE AP_SCHEDULER_CLASS
/*" U2B3.6.4 generated this component from Class AP_SCHEDULER "*/
CONSTANTS
        AP_SCHEDULER
PROPERTIES
        AP_SCHEDULER = 1..n
VARIABLES
        traf_table_index,
        frame_pointer
INVARIANT
        traf_table_index : AP_SCHEDULER --> 0..TrafficTableSize  &
        frame_pointer : AP_SCHEDULER --> 0..4999
INITIALISATION
        traf_table_index :: AP_SCHEDULER --> {0} ||
        frame_pointer :: AP_SCHEDULER --> {0}

OPERATIONS
 design_MAC_frame (thisAP_SCHEDULER) =
BEGIN
   traf_table_index, frame_pointer:(traf_table_index:0..TrafficTableSize &
        traf_table_index<=2+6*(NumberOfMTsPlusBroadcast-1) &
        frame_pointer:0..4999)
  END

END
```

Figure 17-5. The initial B code produced for the AP_SCHEDULER

During the refinement process, appropriate predicates were defined to express the properties of the linking (gluing) invariants between the initial B model and the refined B model. For example, the following invariant is defined in the B code of Figure 17-5:

```
traf_table_index: AP_SCEDULER --> 0..TrafficTableSize
```

which states that variable `traf_table_index` belongs in the range `0..TrafficTableSize`. The specific invariant generates the following proof obligations:

```
1. traf_table_index ≤ TrafficTableSize
2. 0 ≤ traf_table_index
```

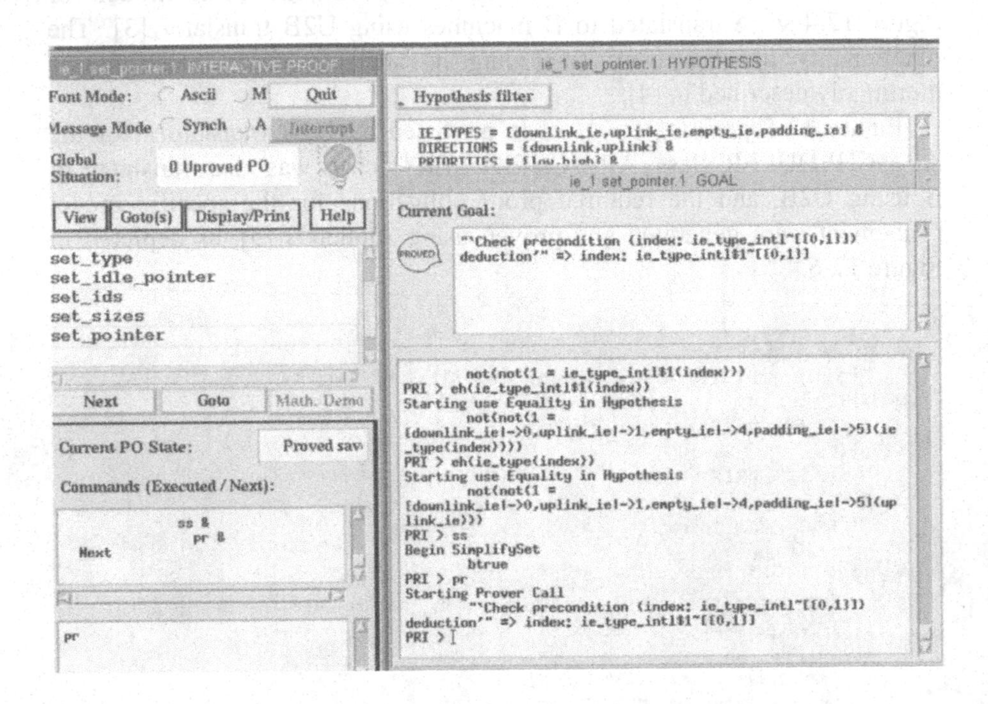

Figure 17-6. The interactive prover of Atelier B

The proof obligations generated were mainly proven using the automatic prover of Atelier B, but there were also cases where the designers had to prove several proof obligations using Atelier B's interactive prover. Moreover, due to the complexity of certain proof obligations, the designers experienced excessive numbers of proof obligations (sometimes more than 100), which were impossible to prove using the interactive prover. In that cases, there were two different alternatives to follow:

- Introduction of an additional intermediate refinement level between successive model refinements, to express some properties at an intermediate level of detail. In many cases this technique can improve the task of proving significantly, while the non proven proof obligations can be proven more easily at the lower refinement levels.
- Introduction of a new module (or more if necessary) to simplify the proving procedure. The new module can make simpler the complex operations (the ones that create excessive numbers of proof obligations). The B machine of the new module is imported to the

lower level refinement with the use of the IMPORTS clause, supported by B language.

In the case study presented, the first alternative was adopted, while in cases of excessive number of proof obligations new intermediate B modules were imported. The refinement process revealed 3.343 proof obligations; 3.106 of them (92,6%) were automatically proven using Atelier B's automatic prover, while 247 proof obligations (7,4%) were proven using the interactive prover of Atelier B.

4. SYSTEM DECOMPOSITION

System decomposition takes place using the decomposition assistant tool [6]. The tool intends to support system partitioning into hardware and software; it accepts formally proven system models described in eventB [7], and based on a profile like the one presented in Figure 17-7, produces subsystems along with the required communication interface. The subsystems and their interfaces can be further formally refined until a fully functional subsystem model is reached. In order to identify and prove the proof obligations required during subsystem refinement, Atelier B is used. The final subsystem implementation emerges through direct translation of the eventB code either to C or to VHDL. For the latter, the BHDL translator [8] developed in the context of PUSSEE Project [9] can be employed.

For the HIPERLAN/2 case study, the B models emerged from the previous design stages were used as input to the decomposition assistant tool in order to partition the system into hardware and software. Part of the system decomposition profile used is presented in Figure 17-7. It contains the number and the names of the subsystems specified by the designers, as well as variable allocation. Communication is deduced from that description, using default communication protocol for accessing data. These communication protocols are likely to be extended.

The system under design was decomposed in two subsystems:

- SS_SCH subsystem which corresponds to the functionality of the system UML-B model of Figure 17-4.
- SS_MODEM subsystem which contains the part of the initial system that contains the HIPERLAN/2 modem functionality.

```
THEORY DECOMPOSITION IS
        SS_SCH={Receive, Transmit, flush, assemble};
        SS_MODEM={IQ_EN, RESET, ENDOP, NOP, CFG, RX}
END
&
THEORY SYNCHRONISATION END
&
THEORY ALLOCATION IS
        Allocate(newi, SS_SCH);
        Allocate(received_pointer, SS_SCH);
        Allocate(source_address, SS_SCH);
        Allocate(destination_address, SS_SCH);
        Allocate(src_ptr, SS_SCH);
        Allocate(current_sar_length, SS_SCH);
        Allocate(dest_length, SS_SCH);
        Allocate(buffer, SS_SCH);
        Allocate(valid_items, SS_SCH);
        Allocate(cell_arrived, SS_SCH);
        Allocate(cell_sent, SS_SCH);
        Allocate(PHY_Mode, SS_MODEM);
        Allocate(Valid, SS_MODEM);
        Allocate(Slot_Num, SS_MODEM);
        Allocate(PDU_Type, SS_MODEM);
        Allocate(EndM, SS_MODEM);
        Allocate(RPT, SS_MODEM);
        Allocate(Datatype, SS_MODEM);
        Allocate(EndT, SS_MODEM);
        Allocate(Enable, SS_MODEM);
        Allocate(EndR, SS_MODEM);
        Allocate(FrameNumber, SS_MODEM)

END
```

Figure 17-7. The decomposition profile for the telecom case study

5. HARDWARE/SOFTWARE ALLOCATION & IMPLEMENTATION

The final step of the design process was the hardware/software allocation and the generation of final code. In the context of the HIPERLAN/2 case study, only the SS_SCH subsystem was implemented. For that purpose, 2291 lines of C code were generated using Atelier B's C code generator and the final C code has been tested on ARM7 TDMI.

6. PUSSEE METHOD EVALUATION

The next paragraphs describe the pros and cons of using the proposed method in the context of an industrial environment for the development of complex telecommunication products.

6.1 Methodology adoption

As already described in Chapter 3, the PUSSEE method relies on the combined use of UML and B. The UML specifications are written in a B compliant manner, using the UML-B profile proposed by the PUSSEE approach.

Even though the PUSSEE method appears to be compatible with the practices used by many telecom companies, the use of B language for proving system properties might be a barrier to the extensive use of the method. The experience gained from the design of HIPERLAN/2 system revealed that a strong mathematical background (especially in the domain of predicate calculus) is required for the engineers that plan to use the PUSSEE approach. As a result, the potential use of the PUSSEE method as part of an existing development process will definitely require training courses of the design teams in order to use productively the proposed approach. The cost/benefit ratio of the latter will definitely play a crucial role in the future adoption of PUSSEE method.

The potential use of PUSSEE method in practice should rely on a combination of the U2B and Atelier B tools. For Atelier B, a possible configuration would include:

- **Atelier B software:** Fully functional, basic configuration for one server including standard graphic user interface, syntax and type checker, proof obligations generator, multi-pass automatic prover and interactive prover with graphic user interface, documentation tools and translators for C and C++.
- **Training:** Training sessions for understanding fundamental principles of the B method (Level 1) and how to develop in B (Level 2).
- **On going support:** Support including, answering questions and determining the best way to efficiently introduce B in a real world design environment development process will be necessary.

The aforementioned requirements reflect the basic version of Atelier B and their cost is 45.000 Euros[1] [5]. Additionally, the configuration may also include:

[1] The prices reported are the official prices of ClearSy S.A., June 2004.

- **User rule proof tools** for validating the mathematical rules added by users during proof, at the cost of 6.000 Euros.
- **Training** for an additional design team at the cost of 15.000 Euros.
- **Four additional licenses** to supplement the basic license at the cost of 3.600 Euros.
- **Maintenance** of the basic tool version including bug fixes and product updates at the cost of 4.800 Euros.

Table 17-1 presents a summary of the total cost of PUSSEE method.

Table 17-1. Total cost of PUSSEE method tools and training

TOOL	COST
U2B translator	Free of charge
Atelier B: Basic configuration	45.000 Euros
User rule proof tools	6.000 Euros
Additional training	15.000 Euros
Additional licenses	3.600 Euros
Maintenance	4.800 Euros
Total cost	74.400 Euros

One additional issue that must also be taken into account is the fact that the method must be mature enough before adopted for the development of commercial products. In the context of a product line, maturity is close related to parameters like stability, on going support, adequate documentation and ability to handle highly complex system models. In its current version, the method is mostly supported by academic tools or prototypes that are still under development. In the context of the case study presented, the tools supporting PUSSEE methodology have proven their value. What remains is to see the combination of PUSSEE methodology and the supporting tools under a more robust development framework.

6.2 Methodology expressiveness

B language is traditionally used for the development of safety critical systems. Thus, in order to provide error free system models of the system under development there are several descriptions that must be imposed. The constraints can be divided into four main methodological notions of B developments:

- *Preservation of the local invariants*

- The operations of a specific machine can be called only by one machine. This restriction prevents data sharing involving multiple write access.
- Simultaneous operation calls are forbidden.
- Each variable of a machine can be altered by, at most, one of the simultaneous substitutions of an operation.

- *Strict tree call structure*
  - Loops within the calling structure of a set of machines are not allowed.
  - Local operations cannot call other operations within the same machine.
- *Encapsulation principle*
  - A variable of a machine can only be written by the operations of the machine containing it.
- *No recursitivity of the operation calls*
  - Simultaneous operation calls are forbidden.

Even though the aforementioned restrictions are essential for the development of formally proven system models, there are cases (especially in the telecom domain) where they might be restrictive. For example, during the design of the AP scheduler a significant part of the scheduler had to be re-designed in order to be compliant with B language primitives. One additional reason that imposed redesigning system parts was the excessive number of proof obligations generated form the initial model. In general, the use of B language requires the definition of a significant number of system properties that must be expressed in the form of invariants. The latter, can lead to significant problems during the proving process, especially when we are dealing with complex systems.

6.3 Tool support

In the context of the HIPERLAN/2 case study, U2B translator, Atelier B and decomposition assistant have been mainly used. The experience gained from their use is described in the next paragraphs.

6.3.1 U2B translator

U2B is a tool, which through a flexible user interface allows translation of UML models (written using the UML-B profile) to B language. It is available in two flavors:

- *U2B3* (version 3) that relies on UML models created using Rational Rose (and the conventions adopted by Rational) and,

- *U2B4* (version 4), which is tool neutral and relies on XML. U2B translates the UML models first to XML language and then to B code. The use of U2B4 for the translation of the UML-B models of the AP scheduler has reveled several restrictions in the way the UML models should be constructed. Additional, there is an inherent difficulty to deal with complicated models.

Additional restrictions of the U2B tool come from the fact that B does not support cyclic structures. As a result, in order to make a B model, the designers had to produce tree-structured UML model. Moreover, during the construction of the UML models the designers should keep in mind that the system will be translated in B and thus using B definitions for variables and functions. If this is not the case, U2B will create a B model, but it will probably not pass the proofing process.

Based on the experience gained for the use of U2B tool, we could say it appears to be a promising tool, which could potentially bridge the gap between UML and B by isolating the designer from the B language details. As a result, designers that are not experts in using B language could use PUSSEE method and take advantage of its benefits. The latter presupposes that UML-B profile and the U2B can be used as a front end that isolates the designers from B language details as much as possible.

6.3.2 Atelier B

Atelier B was the main tool employed for the development of B models of HIPERLAN/2 case study. It was also used for generating and proving the required proof obligations between successive refinements. At the last phases of the development cycle, Atelier B was also used to produce C code for the software part of the final system.

From the total number of POs generated during the refinement process, 92,6% were automatically proven using Atelier B's automatic prover, while only 7,4% of them were proven interactively. Despite the high percentage of automatically proven POs, there were restrictions in Atelier B which are directly related to the nature of B language. In addition, due to the excessive number of proof obligations produced during the early design phases of system design, significant model restructuring was required. Despite the effectiveness of the proving process of Atelier B, there were also cases where the designers came across inefficiencies throughout the proving process. The rule base of Atelier B, in spite of the 2.200 rules it supports, should be enriched in future tool versions in order to ease the proving process. Significant problems were also experienced with proofs that involved cardinal numbers and Σ functions.

Interoperability among the tools is another significant parameter for the design of complex systems. Communication (e.g. model exchange) between

Atelier B with other tools like U2B would definitely be an advantage. Moreover, availability of the tool for different operating systems might be helpful towards this direction.

6.3.3 Decomposition assistant

Decomposition assistant aims at producing system partitions into hardware and software subsystems, based on formally proven system models. In the current version, it only accepts as input system models described in eventB. In the case study presented, this was a major problem since Atelier B relies on B language for generating and proving proof obligations. As a consequence, the designers had to use translators for translating eventB to B and vice versa [10]. This produced significant delays in the partitioning process since the two languages are not fully compatible.

Moreover, the user interface provided by the tool was not adequate, while significant support was required as far as the subsystem interfaces were concerned. What would be expected in forthcoming versions of the tool would be a library of formally proven standard interfaces that could be customized, and possible extended, according to the needs of each subsystem.

6.4 Final product

The last part of the HIPERLAN/2 case study was the generation of C code for the final implementation of the system. The code produced was based on the B implementations constituting the AP scheduler, and for its production Atelier B's automatic translator for C code has been employed. As a general remark we could mention that the code produced by Atelier B was well documented and easy to understand by the designers (the code produced was about 2291 lines in C). Although no optimization techniques were used, it would be preferable to have the ability to produce code for different implementation platforms e.g. for ARM7 TDMI.

In terms of productivity, the code production is a fairly easy process while the fact that system designers are able to produce C code form formally proven to be correct B implementations is definitely an advantage since in allows the detection of design flaws early enough in the design process.

7. SUMMARY

In the previous sections we presented an insight on how PUSSEE method could be employed in a real world design environment. The case study presented is based on HIPERLAN/2 protocol, and has been designed using PUSSEE method and the tools supporting it. For the initial system specification the UML-B profile has been adopted, while for system design B language/method and Atelier B have been used. The latter has been utilized in order to verify formally the correctness of the model refinements in B, as they emerge from the U2B translator.

Based on the experience gained form the design of the HIPERLAN/2 case study, an initial evaluation of PUSSEE method has been presented. In a nutshell, PUSSEE method appears to be a promising design approach, the benefits of which could be exploited in the context of a real world design environment. Nevertheless, there are issues that must be taken into account in future versions of the method, mainly related to the tool interoperability and their efficient use in an existing product development process.

REFERENCES

1. ETSI 2000, *Broadband Radio Access Networks BRAN; HIPERLAN Type 2; Data Link Control (DLC) Layer Part1: Basic Data Transport Functions*, ETSI TS 101 761-1 v1.1.1.
2. C. Snook, M. Walden, *Use of U2B for Specifying B Action Systems*, Proceedings of International Workshop on Refinement of Critical Systems: Methods, Tools and Developments, Grenoble, France, January 2002.
3. C. Snook, M. Butler, *U2B Downloads*, Available at: http://www.ecs.soton.ac.uk/~cfs/U2Bdownloads.htm
4. K. Antonis, N. Voros, *D3.1.2: Specification of the Telecom System-on-Chip Experiment*, IST-2000-30103 PUSSEE, Project Report, 2002.
5. Atelier B, Available at: http://www.Atelier B.societe.com/, 2003.
6. T. Lecomte, *D4.4.1: Methodological Guidelines: Interface based synthesis/ refinement in B*, IST-2000-30103 PUSSEE, Project Report, 2003.
7. ClearSy, *Event B Reference Manual v1.0*, Available at: http://www.atelierb.societe.com/ressources/evt2b/eventb_referencemanual.pdf, 2001.
8. KessDA, *BHDL User Guide Preliminary Version*, Available at: http://www.keesda.com/pussee/bibliography.htm
9. PUSSEE Project, Available at: http://www.keesda.com/pussee, 2003.
10. MATISSE Project, *Event B to B Translator User Manual*, IST-1999-11435, Project Report 1999.

Annex A1

EVALUATION CRITERIA FOR EMBEDDED SYSTEM DESIGN METHODS

Nikolaos S. Voros[1]
Ola Lundkvist[2]
Klaus Kronlöf[3]

[1] INTRACOM S.A., Patra, Greece
[2] Volvo Technology, Göteborg, Sweden
[3] Nokia Research Center, Helsinki, Finland

Abstract: This annex introduces a representative set of criteria for evaluating the applicability of system design methodologies in practice. The criteria cover a wide range of evaluation parameters, as they are perceived by system designers that apply system design methodologies in practice.

Key words: System design methodologies, evaluation criteria.

1. INTRODUCTION

This annex provides details for a common set of evaluation criteria that have been used by industrial case studies for the evaluation of the PUSSEE method [1]. The set of criteria presented constitutes a general evaluation framework that has been instantiated per case study so as to reflect the experience gained, and the problems encountered, during the application of PUSSEE approach in practice. The criteria described are based on the actual problems encountered from system design experts, as they are reported in real world design environments and in literature [2]. They are classified in four main categories:

J. Mermet (ed.), UML-B Specification for Proven Embedded System Design, 293–300.
© 2004 *Kluwer Academic Publishers. Printed in the Netherlands.*

- *Criteria for method adoption* that are related to the usefulness of the proposed method.
- *Expressiveness criteria* that measure how well the design representation (UML-B combination) is suited to express the design concepts of different domains.
- *Tool support criteria* that refer to the level of maturity of the tools supporting the method, and their effective use in practice.
- *Criteria related to the final product*, that describe qualitative characteristics of the final system.

2. CRITERIA FOR METHOD ADOPTION

The criteria belonging in this category describe the methodological requirements of the end users, and focus on the applicability of the PUSSEE approach in practice, as part of an existing design flow. The exact parameters taken into account are described in the sequel.

2.1 Required training

To what extend is training required? (key designers, small team, every design team). Significant parameters for this criterion (related to the technical background of the design teams) are:

- *Familiarity* with the concepts introduced and
- *Method understanding*

The quantification of the aforementioned criterion is based on the number of person hours spent for training and on the total person hours spent in the project. The criterion is calculated as a percentage of the training hours over the total hours of the project.

2.2 Financial cost

What is the cost required for applying the method in practice? Is it justified by the anticipated benefits? It is measured in KEuros and includes the cost of equipment required, plus the cost of training the design teams.

2.3 Compatibility

Is the method compatible with the already existing know how and how easy is to adopt the proposed design technique (partly or totally)? The available values for this parameter are:

- *Incompatible*, meaning that the approach is irrelevant to the existing design practices.
- *Low compatibility*, meaning that the method is partly related to the design process.
- *Medium compatibility*, meaning that several concepts proposed by the method are related to the current design flow.
- *Very compatible*, meaning that most of the method concepts can be applied directly in the current development cycle.
- *Fully compatible* that means the method has one to one correspondence with the existing design stages, and the design practices currently used.

2.4 Maturity

This criterion describes the level of stability of the method, and is based on personal judgement. The values of this parameter are:

- *Immature*: The method is not sufficiently defined.
- *Low maturity*: The method is at least partly defined, and is subjected to frequent or major changes and supplements.
- *Medium maturity*: The method is defined, but is subjected to improvements and minor changes.
- *Mature*: The method is well defined and stable.

2.5 Time spent

This criterion measures total time spent for system design. It emerges as the sum the person hours spent at modelling, design, implementation and integration phases; it reflects the efficiency of the method and is directly influenced by the maturity of the tools supporting the method.

2.6 Support for reuse

It refers to the ability of the method to support component reuse, and incorporate effectively already available components in the proposed design flow. This criterion is measured through a two dimensional matrix that describes the design stages of the method, and the alternative ways of reusing components in each of them. One additional column will indicate the number of components reused per design phase.

2.7 Specification quality improvement

If we detect defects in the informal specifications, we can clearly see that the formalism aided us in improving the specification and possibly removing potential defects from the final product. We classify the defects in the following categories:

- Unclearness
- Ambiguity
- Contradictory
- Missing requirements
- Incorrect requirements

We also classify defects according to severity:

- A *minor defect* means no or insignificant effect on product.
- *Medium defects* mean possibility for noticeable effects on the product.
- *Major defect* means possibility for serious effects on the product or major possibility for noticeable effects.

We count and categorize all defects that are discovered; the severity is mainly based on judgments. The unit of this criterion is a two-dimensional defect vector with the number of defects per type and per severity, and is dependent on the system that is specified. Since there are no instruments that can detect all possible defects, we can only measure defects that we manage to detect.

2.8 Method confidence

The specific criterion reflects how confident are we about the correctness of the result. How foolproof is the method? Is it possible to state incorrect requirements or invariants, or to forget requirements or invariants without discovering them? What assumptions do we need to make in order to produce a system implemented correctly? These are a number of questions concerning how much confident we can be on the correctness of a system developed with the method. Since there is no way to measure this criterion, we can only provide the experience based on personal judgment.

3. EXPRESSIVENESS CRITERIA

Domain related criteria evaluate the effectiveness of the method in terms of appropriateness of the method for the design of systems belonging in a specific domain. The exact criteria are described in the next paragraphs.

3.1 Support for architectural and structural design concepts

They are used to define the partitioning of the system to subsystems and reason about the interactions among the subsystems. Does the method support the functional partitioning approach used? How well is the method suited to express software and hardware architectures that are typical to the particular application at hand? Does the method support the reasoning about the communication between the subsystems (bandwidth, latency, bottlenecks, refinement of communication architecture)? In this case the evaluation profile consists of a two column matrix containing one column with the aforementioned concepts, and a second one describing how they are supported by the proposed approach.

3.2 Support for behavioural design concepts

They are used to define the implementation independent behaviour of the system. These criteria are quite application specific and can be divided into subcategories according to behavioural domains, such as:

- system control behaviour,
- data processing behaviour,
- communication protocol behaviour,
- continuous behaviour, etc.

In the case of heterogeneous systems, it is not enough to be able to express each domain in isolation, but also the interactions of the domains need to be defined. A table containing the behavioural domains and how they are supported by the method used must be provided.

3.3 Description of non functional constraints

These are typically directly derived from system requirements, and are then allocated to subsystems during system partitioning. The *real-time constraints* constitute an important category for many systems; *reliability*, i.e. the fact that a certain functionality is available with very high degree of confidence, and *failure safeness* are also important in several cases. In this case, a two column matrix must provided, describing how non functional constraints are supported and handled per design phase.

3.4 Support for low level design concepts

It reflects the effectiveness of the approach during the system design and implementation phases and takes into account the ability of the method to deal with issues related to software and hardware implementation. For example, is the approach effective for hardware systems? Can be integrated with hardware design processes? How does it support hardware/software integration?

4. TOOL SUPPORT

Tool support evaluation parameters describe aspects related to the effectiveness of the tools used to support the method.

4.1 Maturity

Tool maturity reflects the ability of the tools to be used as part of a production chain. For example (for the tools used by PUSSEE method),

- For *U2B* [3]: To what extent UML is covered for translating UML models to B machines? Is U2B translator tool neutral?
- For *AtelierB* [4]: Are the tool capabilities sufficient for the designers? What is the completeness of the rule base? What is the percentage of the rules needed during system design and not found in the rule base?
- For *eventB decomposition assistant* [5]: Is the decomposition process supported effectively? How well interacts the decomposition tool with AtelierB?
- For *RAVEN* (described in Chapter 10): Is real time constraint checking supported efficiently? Can B/eventB model imported to RAVEN and vive versa? Are there adequate APIs for connecting RAVEN with other tools?
- For *BHDL translator* [6]: What is the actual subset of VHDL supported? Is the produced code synthesizable? Are there any restrictions in the way the eventB models are written?

4.2 User friendliness

This criterion indicates the degree to which the tools can be effectively used by the designers, and is mainly based on the personal opinion of the designers that have experimented with the method the tools supporting it.

4.3 Degree of automation

It describes the ability of the tool to fulfil automatically specific tasks e.g. code generation for hardware/software components, automatic proof of proof obligations etc.

4.4 Interoperability

It is related to the communication among the tools involved in the design process e.g. is it seamless, or does it require model transformations?

4.5 Support of formal proof

Formal proof of the properties of the system under design is the main concept introduced by PUSSEE method. The specific criterion measures how well formal proof is supported per design phase, and to what extent formal proof of real time properties is supported. The specific criterion is close related to the maturity of the tools. It is measured through a two dimensional matrix describing the ability of the tools to support formal proof per design phase.

4.6 Scalability

It describes how scalable are the method and the tool platform, to large size industrial projects. All the aforementioned criteria of this category can contribute on an estimation of the feasibility of the proposed approach to large scale projects. For the evaluation of the specific parameter it is expected to indicate the points of the approach that should be strengthened.

5. CRITERIA RELATED TO FINAL PRODUCT

This set of criteria refers to the qualitative evaluation of the product produced using the method.

5.1 Code quality

Code quality (both for hardware and software) is related to the quality of the final product. The code produced by the tools supporting the method is related to:

- Code understandability
- Code maintainablility
- Ease of integration with existing components
- Error free code
- Compliance with the initial specifications
- Platform neutral code e.g. ANSI C code

5.2 Size of the code

It indicates the size of the produced code, compared to the code produced using the existing design approach (manual code production or code produced by other tools). It is important to mention at this point that the measure of interest is NOT the number of source code lines, but the memory size of the object code and the amount of memory needed for data. In the case of hardware the relevant measures are silicon area, power consumption and execution speed.

5.3 Code optimization

This parameter indicates the ability to produce code optimized according to system requirements.

6. SUMMURY

In the previous sections we presented a set of evaluation criteria for exploring the applicability of the various design methods in practice. The criteria rely on the experience of system designers from three major European companies, and attempt to cover all the design phases from specification down to implementation. The criteria described, constitute a practical evaluation framework that can be applied for evaluating the applicability of every design method.

REFERENCES

1. PUSSEE Project. Available at: http://www.keesda.com/pussee, 2004.
2. P. Cavalloro, C. Gendarme, K. Kronlöf, J. Mermet, J. van Sas, K. Tiensyrjä and N. S. Voros, *System Level Design with Reuse of System IP*, Kluwer Academic Publishers, 2003.
3. C. Snook, M. Butler, *U2B Downloads*, Available at: http://www.ecs.soton.ac.uk/~cfs/U2Bdownloads.htm.
4. Atelier B, Available at: http://www.Atelier B.societe.com/, 2003.
5. T. Lecomte, *D4.4.1: Methodological Guidelines: Interface based synthesis/ refinement in B*, IST-2000-30103 PUSSEE, Project Report, 2003.
6. KeesDA, *BHDL User Guide. Preliminary Version*, Available at http://www.keesda.com.